BUFFON'S NATURAL HISTORY
(VOLUME I)

LECTOR HOUSE PUBLIC DOMAIN WORKS

BUFFON'S NATURAL HISTORY (VOLUME I)

GEORGES LOUIS LECLERC DE BUFFON (COMTE DE BUFFON), JAMES SMITH BARR

ISBN: 978-93-89560-30-5

Published: 1807

LECTOR HOUSE LLP
E-MAIL: lectorpublishing@gmail.com

BUFFON'S NATURAL HISTORY.

CONTAINING A THEORY OF THE EARTH, A GENERAL HISTORY OF MAN, OF THE BRUTE CREATION, AND OF VEGE TABLES, MINERALS, &C. &C.

FROM THE FRENCH.
WITH NOTES BY THE TRANSLATOR.

BY

GEORGES LOUIS LECLERC DE BUFFON

TRANSLATED BY
JAMES SMITH BARR

IN TEN VOLUMES.
VOL. I.

1807.

PREFACE.

We should certainly be guilty of a gross absurdity if, in an age like the present, we were to enter into an elaborate discussion on the advantages to be derived from the study of NATURAL HISTORY; the ancients recommended it as useful, instructive, and entertaining; and the moderns have so far pursued and cultivated this first of sciences, that it is now admitted to be the source of universal instruction and knowledge; where every active mind may find subjects to amuse and delight, and the artist a never failing field to enrich his glowing imagination.

It would have been singular if, on such a subject, a number of authors had not submitted the produce of their observations and labour; many have written upon Natural Philosophy, but the Comte de BUFFON stands eminently distinguished among them; he has entered into a minute investigation, and drawn numberless facts from unwearied observations far beyond any other, and this he has accomplished in a style fully accordant with the importance of his subject. Ray, Linnæus, Rheaumur, and other of his cotemporaries, deserve much credit for their classing of animals, vegetables, &c. but it was BUFFON alone who entered into a description of their nature, habits, uses, and properties. In his Theory of the Earth he has displayed a wonderful ingenuity, and shewn the general order of Nature with a masterly hand, although he may be subject to some objections for preferring physical reasonings on general causes, rather than allowing aught to have arisen from supernatural agency, or the will of the Almighty. In this he has followed the example of all great philosophers, who seem unwilling to admit that the formation of any part of the Universe is beyond their comprehension.

As the works of this Author will best speak for themselves, we shall avoid unnecessary panegyric, hoping they will have received no material injury in the following translation; we shall therefore content ourselves with observing, that in our plan we have followed that adopted by the Comte himself in a latter edition, from which he exploded his long and minute treatises on anatomy and mensuration; though elegant and highly finished in themselves, they appeared to us of too abstruse and confined a nature for general estimation, and which we could not have gone into without almost doubling the expence; a circumstance we had to guard against, for the advantage of those of our readers to whom that part would have been totally uninteresting.

As to this edition, we presume it is no vain boast, that every exertion has been made to do justice to a work of such acknowledged merit. In the literary part, it has been the Proprietor's chief endeavour to preserve the spirit and accuracy of the Author, as far as could be done in translating from one language into another; and it is with gratitude he acknowledges, that those endeavours have been amply supported by the engraver; for the decorative executions of MILTON will remain a lasting monument of his abilities, as long as delicacy in the arts is held in estimation.

CONTENTS

Page

VOLUME I

PROOF OF THE THEORY OF THE EARTH.

ARTICLE I.

ARTICLE II.

ARTICLE III.

ARTICLE IV.

ARTICLE V.

ARTICLE VI.

ARTICLE VII.

ARTICLE VIII.

ARTICLE IX.

ARTICLE X.

BUFFON'S NATURAL HISTORY.

VOLUME I

THE THEORY OF THE EARTH.

Neither the figure of the earth, its motion, nor its external connections with the rest of the universe, pertain to our present investigation. It is the internal structure of the globe, its composition, form, and manner of existence which we purpose to examine. The general history of the earth should doubtless precede that of its productions, as a necessary study for those who wish to be acquainted with Nature in her variety of shapes, and the detail of facts relative to the life and manners of animals, or to the culture and vegetation of plants, belong not, perhaps, so much to Natural History, as to the general deductions drawn from the observations that have been made upon the different materials which compose the terrestrial globe: as the heights, depths, and inequalities of its form; the motion of the sea, the direction of mountains, the situation of rocks and quarries, the rapidity and effects of currents in the ocean, &c. This is the history of nature in its most ample extent, and these are the operations by which every other effect is influenced and produced. The theory of these effects constitutes what may be termed a primary science, upon which the exact knowledge of particular appearances as well as terrestrial substances entirely depends. This description of science may fairly be considered as appertaining to physics; but does not all physical knowledge, in which no system is admitted, form part of the History of Nature?

In a subject of great magnitude, whose relative connections are difficult to trace, and where some facts are but partially known, and others uncertain and obscure, it is more easy to form a visionary system, than to establish a rational theory; thus it is that the Theory of the Earth has only hitherto been treated in a vague and hypothetical manner; I shall therefore but slightly mention the singular notions of some authors who have written upon the subject.

The first hypothesis I shall allude to, deserves to be mentioned more for its ingenuity than its reasonable solidity; it is that of an English astronomer, (WHISTON) versed in the system of NEWTON, and an enthusiastic admirer of his philosophy; convinced that every event which happens on the terrestrial globe, depends upon the motions of the stars, he endeavours to prove, by the assistance of mathematical calculations, that the tail of a comet has produced every alteration the earth has ever undergone.

The next is the formation of an heterodox theologician, (BURNET) whose brain was so heated with poetical visions, that he imagined he had seen the creation of the universe. After explaining what the earth was in its primary state, when it sprung from nothing; what changes were occasioned by the deluge; what it has been and what it is, he then assumes a prophetic style, and predicts what will be its state after the destruction of the human race.

The third comes from a writer (WOODWARD) certainly a better and more extensive observer of nature than the two former, though little less irregular and confused in his ideas; he explains the principal appearances of the globe, by an immense abyss in the bowels of the earth, which in his opinion is nothing more than a thin crust that serves as a covering to the fluid it incloses.

The whole of these hypotheses are raised on unstable foundations; have given no light upon the subject, the ideas being unconnected, the facts confused, and the whole confounded with a mixture of physic and fable; and consequently have been adopted only by those who implicitly believe opinions without investigation, and who, incapable of distinguishing probability, are more impressed with the wonders of the marvellous than the relation of truth.

What we shall say on this subject will doubtless be less extraordinary, and appear unimportant, if put in comparison with the grand systems just mentioned, but it should be remembered that it is an historian's business to describe, not invent; that no suppositions should be admitted upon subjects that depend upon facts and observation; that his imagination ought only to be exercised for the purpose of combining observations, rendering facts more general, and forming one connected whole, so as to present to the mind a distinct arrangement of clear ideas and probable conjectures; I say probable, because we must not expect to give exact demonstration on this subject, that being confined to mathematical sciences, while our knowledge in physics and natural history depends solely upon experience, and is confined to reasoning upon inductions.

In the history of the Earth, we shall therefore begin with those facts that have been obtained from the experience of time, together with what we have collected by our own observations.

This immense globe exhibits upon its surface heights, depths, plains, seas, lakes, marshes, rivers, caverns, gulphs, and volcanos; and upon the first view of these objects we cannot discover in their dispositions either order or regularity. If we penetrate into its internal part, we shall there find metals, minerals, stones, bitumens, sands, earths, waters, and matters of every kind, placed as it were by chance, and without the smallest apparent design. Examining with a more strict attention, we discover sunk mountains, caverns filled, rocks split and broken, countries swallowed up, and new islands rising from the ocean; we shall also perceive heavy substances placed above light ones, hard bodies surrounded with soft; in short, we shall there find matter in every form, wet and dry, hot and cold, solid and brittle, mixed in such a sort of confusion as to leave room to compare them only to a mass of rubbish and the ruins of a wrecked world.

We inhabit these ruins however with a perfect security. The various generations of men, animals, and plants, succeed each other without interruption; the earth produces fully sufficient for their subsistence; the sea has its limits; its motions and the currents of air are regulated by fixed laws: the returns of the seasons are certain and regular; the severity of the winter being constantly succeeded by the beauties of the spring: every thing appears in order, and the earth, formerly a CHAOS, is now a tranquil and delightful abode, where all is animated, and regulated by such an amazing display of power and intelligence as fills us with admiration, and elevates our minds with the most sublime ideas of an all-potent and wonderful Creator.

Let us not then draw any hasty conclusions upon the irregularities of the surface of the earth, nor the apparent disorders in the interior parts, for we shall soon discover the utility, and even the necessity of them; and, by considering them with a little attention, we shall, perhaps, find an order of which we had no conception, and a general connection that we could neither perceive nor comprehend, by a slight examination: but in fact, our knowledge on this subject must always be confined. There are many parts of the surface of the globe with which we are entirely unacquainted, and have but partial ideas of the bottom of the sea, which in many places we have not been able to fathom. We can only penetrate into the coat of the earth; the greatest caverns and the deepest mines do not descend above the eight thousandth part of its diameter, we can therefore judge only of the external and mere superficial part; we know, indeed, that bulk for bulk the earth weighs four times heavier than the sun, and we also know the proportion its weight bears with other planets; but this is merely a relative estimation; we have no certain standard nor proportion; we are so entirely ignorant of the real weight of the materials, that the internal part of the globe may be a void space, or composed of matter a thousand times heavier than gold; nor is there any method to make further discoveries on this subject; and it is with the greatest difficulty any rational conjectures can be formed thereon.

We must therefore confine ourselves to a correct examination and description of the surface of the earth, and to those trifling depths to which we have been enabled to penetrate. The first object which presents itself is that immense quantity of water which covers the greatest part of the globe; this water always occupies the lowest ground, its surface always level, and constantly tending to equilibrium and rest; nevertheless it is kept in perpetual agitation by a powerful agent, which opposing its natural tranquillity, impresses it with a regular periodical motion, alternately raising and depressing its waves, producing a vibration in the total mass, by disturbing the whole body to the greatest depths. This motion we know has existed from the commencement of time, and will continue as long as the sun and moon, which are the causes of it.

By an examination of the bottom of the sea, we discover that to be fully as irregular as the surface of the earth; we there find hills and vallies, plains and cavities, rocks and soils of every kind: we there perceive that islands are only the summits of vast mountains, whose foundations are at the bottom of the Ocean; we also find other mountains whose tops are nearly on a level with the surface of the water, and rapid currents which run contrary to the general movement: they sometimes run in the same direction, at others, their motions are retrograde, but never exceeding their bounds, which appear to be as fixed and invariable as those which confine the rivers of the earth. In one part we meet with tempestuous regions, where the winds blow with irresistible fury, where the sea and the heavens equally agitated, join in contact with each other, are mixed and confounded in the general shock:

in others, violent intestine motions, tumultuous swellings, water-spouts, and extraordinary agitations, caused by volcanos, whose mouths though a considerable depth under water, yet vomit fire from the midst of the waves, and send up to the clouds a thick vapour, composed of water, sulphur, and bitumen. Further we perceive dreadful gulphs or whirlpools, which seem to attract vessels, merely to swallow them up. On the other hand, we discover immense regions, totally opposite in their natures, always calm and tranquil, yet equally dangerous; where the winds never exert their power, where the art of the mariner becomes useless, and where the becalmed voyager must remain until death relieves him from the horrors of despair. In conclusion, if we turn our eyes towards the northern or southern extremities of the globe, we there perceive enormous flakes of ice separating themselves from the polar regions, advancing like huge mountains into the more temperate climes, where they dissolve and are lost to the sight.

Exclusive of these principal objects the vast empire of the sea abounds with animated beings, almost innumerable in numbers and variety. Some of them, covered with light scales, move with astonishing celerity; others, loaded with thick shells, drag heavily along, leaving their track in the sand; on others Nature has bestowed fins, resembling wings, with which they raise and support themselves in the air, and fly to considerable distances; while there are those to whom all motion has been denied, who live and die immoveably fixed to the same rock: every species, however, find abundance of food in this their native element. The bottom of the sea, and the shelving sides of the various rocks, produce great abundance of plants and mosses of different kinds; its soil is composed of sand, gravel, rocks, and shells; in some parts a fine clay, in others a solid earth, and in general it has a complete resemblance to the land which we inhabit.

Let us now take a view of the earth. What prodigious differences do we find in different climates? What a variety of soils? What inequalities in the surface? but upon a minute and attentive observation we shall find the greatest chain of mountains are nearer the equator than the poles; that in the Old Continent their direction is more from the east to west than from the north to south; and that, on the contrary, in the New World they extend more from north to south than from east to west; but what is still more remarkable, the form and direction of those mountains, whose appearance is so very irregular, correspond so precisely, that the prominent angles of one mountain are always opposite to the concave angles of the neighbouring mountain, and are of equal dimensions, whether they are separated by a small valley or an extensive plain. I have also observed that opposite hills are nearly of the same height, and that, in general, mountains occupy the middle of continents, islands, and promontories, which they divide by the greatest lengths.

In following the courses of the principal rivers, I have likewise found that they are almost always perpendicular with those of the sea into which they empty themselves; and that in the greatest part of their courses they proceed nearly in the direction of the mountains from which they derive their source.

The sea shores are generally bounded with rocks, marble, and other hard stones, or by earth and sand which has accumulated by the waters from the sea, or been brought down by the rivers; and I observe that opposite coasts, separated only by an arm of the sea, are composed of similar materials, and the beds of the earth are exactly the same. Volcanos I find exist only in the highest mountains; that many of them are entirely extinct; that some are connected with others by subterraneous passages, and that their explosions frequently happen at one and the same time. There are similar correspondences between certain lakes and neighbouring seas; some rivers suddenly disappear, and seem to precipitate themselves into the earth. We also find internal, or mediterranean seas, constantly receiving an enormous quantity of water from a number of rivers without ever extending their bounds, most probably discharging by subterraneous passages all their superfluous supplies. Lands which have been long inhabited are easily distinguished from those new countries where the soil appears in a rude state, where the rivers are full of cataracts, where the earth is either overflowed with water, or parched up with drought, and where every spot upon which a tree will grow is covered with uncultivated woods.

Pursuing our examination in a more extensive view, we find that the upper strata that surrounds the globe, is universally the same. That this substance which serves for the growth and nourishment of animals and vegetables, is nothing but a composition of decayed animal and vegetable bodies reduced into such small particles, that their former organization is not distinguishable; or penetrating a little further, we find the real earth, beds of sand, lime-stone, argol, shells, marble, gravel, chalk, &c. These beds are always parallel to each other and of the same thickness throughout their whole extent. In neighbouring hills beds of the same materials are invariably found upon the same levels, though the hills are separated by deep and extensive intervals. All beds of earth, even the most solid strata, as rocks, quarries of marble, &c. are uniformly divided by perpendicular fissures; it is the same in the largest as well as smallest depths, and appears a rule which nature invariably pursues.

In the very bowels of the earth, on the tops of mountains, and even the most remote parts from the sea, shells, skeletons of fish, marine plants, &c. are frequently found, and these shells, fish, and plants, are exactly similar to those which exist in the Ocean. There are a prodigious quantity of petrified shells to be met with in an infinity of places, not only inclosed in rocks, masses of marble, lime-stone, as well as in earth and clays, but are actually incorporated and filled with the very substance which surrounds them. In short, I find myself convinced, by repeated observations, that marbles, stones, chalks, marls, clay, sand, and almost all terrestrial substances, wherever they may be placed, are filled with shells and other substances, the productions of the sea.

These facts being enumerated, let us now see what reasonable conclusions are to be drawn from them.

The changes and alterations which have happened to the earth, in the space of the last two or three thousand years, are very inconsiderable indeed, when compared with those important revolutions which must have taken place in those ages which immediately followed the creation; for as all terrestrial substances could only acquire solidity by the continued action of gravity, it would be easy to demonstrate that the surface of the earth was much softer at first than it is at present, and consequently the same causes which now produce but slight and almost imperceptible changes during many ages, would then effect great revolutions in a very short space. It appears to be a certain fact, that the earth which we now inhabit, and even the tops of the highest mountains, were formerly covered with the sea, for shells and other marine productions are frequently found in almost every part; it appears also that the water remained a considerable time on the surface of the earth, since in many places there have been discovered such prodigious banks of shells, that it is impossible so great a multitude of animals could exist at the same time: this fact seems likewise to prove, that although the materials which composed the surface of the earth were then in a state of softness, that rendered them easy to be disunited, moved and transported by the waters, yet that these removals were not made at once; they must indeed have been successive, gradual, and by degrees, because these kind of sea productions are frequently met with more than a thousand feet below the surface, and such a considerable thickness of earth and stone could not have accumulated but by the length of time. If we were to suppose that at the Deluge all the shell-fish were raised from the bottom of the sea, and transported over all the earth; besides the difficulty of establishing this supposition, it is evident, that as we find shells incorporated in marble and in the rocks of the highest mountains, we must likewise suppose that all these marbles and rocks were formed at the same time, and that too at the very instant of the Deluge; and besides, that previous to this great revolution there were neither mountains, marble, nor rocks, nor clays, nor matters of any kind similar to those we are at present acquainted with, as they almost all contain shells and other productions of the sea. Besides, at the time of the Deluge, the earth must have acquired a considerable degree of solidity, from the action of gravity for more than sixteen centuries, and consequently it does not appear possible that the waters, during the short time the Deluge lasted, should have overturned and dissolved its surface to the greatest depths we have since been enabled to penetrate.

But without dwelling longer on this point, which shall hereafter be more amply discussed, I shall confine myself to well-known observations and established facts. There is no doubt but that the waters of the sea at some period covered and remained for ages upon that part of the globe which is now known to be dry land; and consequently the whole continents of Asia, Europe, Africa, and America, were then the bottom of an ocean abounding with similar productions to those which the sea at present contains: it is equally certain that the different strata which compose the earth are parallel and horizontal, and it is evident their being in this situation is the operation of the waters which have collected and accumulated by degrees the different materials, and given them the same position as the water itself always assumes. We observe that the position of strata is almost universally horizontal: in plains it is exactly so, and it is only in the mountains that they are inclined to the horizon, from their having been originally formed by a sediment deposited upon an inclined base. Now I insist that these strata must have been formed by degrees, and not all at once, by any revolution whatever, because strata, composed of heavy materials, are very frequently found placed above light ones, which could not be, if, as some authors assert, the whole had been mixed with the waters at the time of the Deluge, and afterwards precipitated; in that case every thing must have had a very different appearance to that which now exists. The heaviest bodies would have descended first, and each particular stratum would have been arranged according to its weight and specific gravity, and we should not see solid rocks or metals placed above light sand any more than clay under coal.

We should also pay attention to another circumstance; it confirms what we have said on the formation of the strata; no other cause than the motions and sediments of water could possibly produce so regular a position of it, for the highest mountains are composed of parallel strata as well as the lowest plains, and therefore we cannot attribute the origin and formation of mountains to the shocks of earthquakes, or eruptions of volcanos. The small eminences which are sometimes raised by volcanos, or convulsive motions of the earth, are not by any means composed of parallel strata, they are a mere disordered heap of matters thrown confusedly together; but the horizontal and parallel position of the strata must necessarily proceed from the operations of a constant cause and motion, always regulated and directed in the same uniform manner.

From repeated observations, and these incontrovertible facts, we are convinced that the dry part of the globe, which is now habitable, has remained for a long time under the waters of the sea, and consequently this earth underwent the same fluctuations and changes which the bottom of the ocean is at present actually undergoing. To discover therefore what formerly passed on the earth, let us examine what now passes at the bottom of the sea, and from thence we shall soon be enabled to draw rational conclusions with regard to the external form and internal composition of that which we inhabit.

From the Creation the sea has constantly been subject to a regular flux and reflux: this motion, which raises and falls the waters twice in every twenty-four hours, is principally occasioned by the action of the moon, and is much greater under the equator than in any other climates. The earth performs a rapid motion on its axis, and consequently has a centrifugal force, which is also the greatest at the equator; this latter, independent of actual observation, proves that the earth is not perfectly spherical, but that it must be more elevated under the equator then at the poles.

From these combined causes, the ebbing and flowing of the tides, and the motion of the earth, we may fairly conclude, that although the earth was a perfect sphere in its original form, yet its diurnal motion, together with the constant flux and reflux of the sea, must, by degrees, in the course of time, have raised the equatorial parts, by carrying mud, earth, sand, shells, &c. from other climes, and there depositing of them. Agreeable to this idea the greatest irregularities must be found, and, in fact, are found near the equator. Besides, as this motion of the tides is made by diurnal alternatives, and been repeated, without interruption, from the commencement of time, is it not natural to imagine, that each time the tide flows the water carries a small quantity of matter from one place to another, which may fall to the bottom like a sediment, and form those parallel and horizontal strata which are every where to be met with? for the whole motion of the water, in the flux and reflux, being horizontal, the matters carried away with them will naturally be deposited in the same parallel direction.

But to this it may be said, that as the flux and reflux of the waters are equal and regularly succeed, two motions would counterpoise each other, and the matters brought by the flux would be returned by the reflux, and of course this cause for the formation of the strata must be chimerical; that the bottom of the sea could not experience any material alteration by two uniform motions, wherein the effects of the one would be regularly destroyed by the other; much less could they change the original form by the production of heights and inequalities.

To which it may be answered, that the alternate motions of the waters are not equal, the sea having a constant motion from the east to the west, besides, the agitation, caused by the winds, opposes and prevents the equality of the tides. It will also be admitted, that by every motion of which the sea is susceptible, particles of earth and other matters will be carried from one place and deposited in another; and these collections will necessarily assume the form of horizontal and parallel strata, from the various combinations of the motions of the sea always tending to move the earth, and to level these materials wherever they fall, in the form of a sediment. But this objection is easily obviated by the well-known fact, that upon all coasts, bordering the sea, where the ebbing and flowing of the tide is observed, the flux constantly brings in a number of things which the reflux does not carry back. There are many places upon which the sea insensibly gains and gradually covers over, while there are others from which it recedes, narrowing as it were its limits, by depositing earth, sands, shells, &c. which naturally take an horizontal position; these matters accumulate by degrees in the course of time, and being raised to a certain point gradually exclude the water, and so become part of the dry land for ever after.

But not to leave any doubt upon this important point, let us strictly examine into the possibility of a mountain's being formed at the bottom of the sea by the motions and sediments of the waters. It is certain that on a coast which the sea beats with violence during the agitation of its flow, that every wave must carry off some part of the earth; for wherever the sea is bounded by rocks, it is a plain fact that the water by degrees wears away those rocks, and consequently carries away small particles every time the waves retire; these particles of earth and stone will necessarily be transported to some distance, and being arrived where the agitation of the water is abated, and left to their own weight, they precipitate to the bottom in form of a sediment, and there form a first stratum, either horizontal or inclined, according to the position of the surface upon which they fall; this will shortly be covered by a similar stratum produced by the same cause, and thus will a considerable quantity of matter be almost insensibly collected together, and the strata of which will be placed, parallel to each other.

This mass will continue to increase by new sediments, and by gradually accumulating, in the course of time become a mountain at the bottom of the sea, exactly similar to those we see on dry land, both as to outward form and internal composition. If there happen to be shells in this part of the sea, where we have supposed this deposit to be made, they will be filled and covered with the sediment, and incorporated in the deposited matter, making a part of the whole mass, and they will be found situated in the parts of the mountain according to the time they had been there deposited; those that lay at the bottom, previous to the formation of the first stratum, will be found in the lowest, and so according to the time of their being deposited, the latest in the most elevated parts.

So likewise, when the bottom of the sea, at particular places, is troubled by the agitation of the water, there will necessarily ensue, in the same manner, a removal of earth, shells, and other matters, from the troubled to other parts; for we are assumed by all divers, that at the greatest depths they descend, i. e. twenty fathoms, the bottom of the sea is so troubled by the agitation of the waters, that the mud and shells are carried to considerable distances, consequently transportations of this kind are made in every part of the sea, and this matter falling must form eminences, composed like our mountains, and in every respect similar; therefore the flux and reflux, by the winds, the currents, and all the motions of the water, must inevitably create inequalities at the bottom of the sea.

Nor must we imagine that these matters cannot be transported to great distances, because we daily see grain, and other productions of the East and West Indies, arriving on our own coasts.[1] It is true these bodies are specifically lighter than water, whereas the substances of which we have been speaking are specifically heavier; but, however, being reduced to an impalpable powder, they may be sustained a long time in the water so as to be conveyed to considerable distances.

It has been supposed that the sea is not troubled at the bottom, especially if it is very deep, by the agitations produced by the winds and tides; but it should be recollected that the whole mass, however deep, is put in motion by the tides, and that in a liquid globe this motion would be communicated to the very centre; that the power which produces the flux and reflux is a penetrating force, which acts proportionably upon every particle of its mass, so that we can determine by calculation the quantity of its force at different depths; but, in short, this point is so certain, that it cannot be contested but by refusing the evidence of reason.

Therefore, we cannot possibly have the least doubt that the tides, the winds, and every other cause which agitates the sea, must produce eminences and inequalities at the bottom, and those heights must ever be composed of horizontal or equally inclined strata. These eminences will gradually encrease until they become hills, which will rise in situations similar to the waves that produce them; and if there is a long extent of soil, they will continue to augment by degrees; so that in course of time they will form a vast chain of mountains. Being formed into mountains, they become an obstacle to and interrupt the common motion of the sea, producing at the same time other motions, which are generally called currents. Between two neighbouring heights at the bottom of the sea a current will necessarily be formed, which will follow their common direction, and, like a river, form a channel, whose angles will be alternately opposite during the whole extent of its course. These heights will be continually increasing, being subject only to the motion of the flux, for the waters during the flow will leave the common sediment upon their ridges; and those waters which are impelled by the current will force

[1] Particularly Scotland and Ireland.

along with them, to great distances, those matters which would be deposited between both, at the same time hollowing out a valley with corresponding angles at their foundation. By the effects of these motions and sediments the bottom of the sea, although originally smooth, must become unequal, and abounding with hills and chains of mountains, as we find it at present. The soft materials of which the eminences are originally composed will harden by degrees with their own weight; some forming parts, purely angular, produce hills of clay; others, consisting of sandy and crystalline particles, compose those enormous masses of rock and flint from whence crystal and other precious stones are extracted; those formed with stony particles, mixed with shells, form those of lime-stone and marble, wherein we daily meet with shells incorporated; and others, compounded of matter more shelly, united with pure earth, compose all our beds of marle and chalk. All these substances are placed in regular beds, and all contain heterogeneous matter; marine productions are found among them in abundance, and nearly according to the relation of their specific weights; the lightest shells in chalk, and the heaviest in clay and lime-stone; these shells are invariably filled with the matter in which they have been inclosed, whether stones or earth; an incontestible proof that they have been transported with the matter that fills and surrounds them, and that this matter was at that time in an impalpable powder. In short, all those substances whose horizontal situations have been established by the level of the waters of the sea, will constantly preserve their original position.

But here it may be observed, that most hills, whose summits consist of solid rocks, stone, or marble, are formed upon small eminences of much lighter materials, such for instance as clay, or strata of sand, which we commonly find extended over the neighbouring plains, upon which it may be asked, how, if the foregoing theory be just, this seemingly contradictory arrangement happens? To me this phenomenon appears to be very easy and naturally explained. The water at first acts upon the upper stratum of coasts, or bottom of the sea, which commonly consists of clay or sand, and having transported this, and deposited the sediment, it of course composes small eminences, which form a base for the more heavy particles to rest upon. Having removed the lighter substances, it operates upon the more heavy, and by constant attrition reduces them to an impalpable powder; which it conveys to the same spot, and where, being deposited, these stony particles, in the course of time, form those solid rocks and quarries which we now find upon the tops of hills and mountains. It is not unlikely that as these particles are much heavier than sand or clay, that they were formerly a considerable depth under a strata of that kind, and now owe their high situations to having been last raised up and transported by the motion of the water.

To confirm what we here assert, let us more closely investigate the situation of those materials which compose the superficial outer part of the globe, indeed the only part with which we have any knowledge. The different beds of strata in stone quarries are almost all horizontal, or regularly inclined; those whose foundations are on clays or other solid matters are clearly horizontal, especially in plains. The quarries wherein we find flint, or brownish grey free-stone, in detached portions, have a less regular position, but even in those the uniformity of nature plainly appears, for the horizontal or regularly inclined strata are apparent in quarries where these stones are found in great masses. This position is universal, except in quarries where flint and brown free-stone are found in small detached portions, the formation of which we shall prove to have been posterior to those we have just been treating of; for granite, vitrifiable sand, argol, marble, calcareous stone, chalk, and marles, are always deposited in parallel strata, horizontally or equally inclined; the original formation of these are easily discovered, for the strata are exactly horizontal and very thin, and are arranged above each other like the leaves of a book. Beds of sand, soft and hard clay, chalk, and shells, are also either horizontal or regularly inclined. Strata of every kind preserves the same thickness throughout its whole extent, which often occupies the space of many miles, and may be traced still farther by close and exact observations. In a word, the materials of the globe, as far as mankind have been enabled to penetrate, are arranged in an uniform position, and are exactly similar.

The strata of sand and gravel which have been washed down from mountains must in some measure be excepted; in vallies they are sometimes of a considerable extent, and are generally placed under the first strata of the earth; in plains, they are as even as the most ancient and interior strata, but near the bottom and upon the ridges of hills they are inclined, and follow the inclination of the ground upon which they have flowed. These being formed by rivers and rivulets, which are constantly in vallies changing their beds, and dragging these sands and gravel with them, they are of course very numerous. A small rivulet flowing from the neighbouring heights, in the course of time will be sufficient to cover a very spacious valley with a strata of sand and gravel, and I have often observed in hilly countries, whose base, as well as the upper stratum, was hard clay, that above the source of the rivulet the clay is found immediately under the vegetable soil, and below it there is the thickness of a foot of sand upon the clay, and which extends itself to a considerable distance. These strata formed by rivers are not very ancient, and are easily discovered by the inequality of their thickness, which is constantly varying, while the ancient strata preserves the same dimensions throughout; they are also to be known by the matter itself, which bears evident marks of having been smoothed and rounded by the motions of the water. The same may be said of the turf and perished vegetables which are found below the first stratum of earth in marshy grounds; they cannot be considered as ancient, but entirely produced by successive heaps of decayed trees and other plants. Nor are the strata of slime and mud, which are found in many countries, to be considered as ancient productions, having been formed by stagnated waters or inundations of rivers, and are neither so horizontal, nor equally inclined, as the strata anciently produced by the regular motions of the sea. In the strata formed by rivers we constantly meet with river, but scarcely ever sea shells, and the few that are found are broken and irregularly placed; whereas in the ancient strata there are no river shells; the sea shells are in great quantities, well preserved, and all placed in the same manner, having been transported at the same time and by the same cause. How are we to account for this astonishing regularity? Instead of regular strata, why do we not meet with the matters that compose the earth jumbled together, without any kind of order? Why are not rocks, marbles, clays, marles, &c. variously dispersed, or joined by irregular or vertical strata? Why are not the heaviest bodies uniformly found placed beneath the lightest? It is easy to perceive that this uniformity of nature, this organization of earth, this connection of different materials, by parallel strata, without respect to their weights, could only be produced by a cause as powerful and constant as the motion of the sea, whether occasioned by the regular winds or by that of the flux and reflux, &c.

These causes act with greater force under the equator than in other climates, for there the winds are more regular and the tides run higher; the most extensive chains of mountains are also near the equator. The mountains of Africa and Peru are the highest known, they frequently extend themselves through whole provinces, and stretch, to considerable distances under the ocean. The mountains of Europe and Asia, which extend from Spain to China, are not so high as those of South America and Africa. The mountains of the North, according to the relation of travellers, are only hills in comparison with those of the Southern countries. Besides, there are very few islands in the Northern Seas, whereas in the torrid zone they are almost innumerable, and as islands are only the summits of mountains, it is evident that the surface of the earth has many more inequalities towards the equator than in the northerly climes.

It is therefore evident that the prodigious chain of mountains which run from the West to the East in the old continent, and from the North to the South in the new, must have been produced by the general motion of the tides; but the origin of all the inferior mountains must be attributed to the particular motions of currents, occasioned by the winds and other irregular agitations of the sea: they may probably have been produced by a combination of all those motions, which must be capable of infinite variations, since the winds and different positions of islands and coasts change the regular course of the tides, and compel them to flow in every possible direction: it is, therefore, not in the least astonishing that we should see considerable eminences, whose courses have no determined direction. But it is sufficient for our present purpose to have demonstrated that mountains are not the produce of earthquakes, or other accidental causes, but that they are the effects resulting from the general order of nature, both as to their organization and the position of the materials of which they are composed.

But how has it happened that this earth which we and our ancestors have inhabited for ages, which, from time immemorial, has been an immense continent, dry and removed from the reach of the waters, should, if formerly the bottom of the ocean, be actually larger than all the waters, and raised to such a height as to be distinctly separated from them? Having remained so long on the earth, why have the waters now abandoned it? What accident, what cause could produce so great a change? Is it possible to conceive one possessed of sufficient power to produce such an amazing effect?

These questions are difficult to be resolved, but as the facts are certain and incontrovertible, the exact manner in which they happened may remain unknown, without prejudicing the conclusions that may be drawn from them; nevertheless, by a little reflection, we shall find at least plausible reasons for these changes. We daily observe the sea gaining ground on some coasts and losing it on others; we know that the ocean has a continued regular motion from East to West; that it makes loud and violent efforts against the low lands and rocks which confine it; that there are whole provinces which human industry can hardly secure from the rage of the sea; that there are instances of islands rising above, and others being sunk under the waters. History speaks of much greater deluges and inundations. Ought not this to incline us to believe that the surface of the earth has undergone great revolutions, and that the sea may have quitted the greatest part of the earth which it formerly covered? Let us but suppose that the old and new worlds were formerly but one continent, and that the Atlantis of Plato was sunk by a violent earthquake; the natural consequence would be, that the sea would necessarily have flowed in from all sides, and formed what is now called the Atlantic Ocean, leaving vast continents dry, and possibly those which we now inhabit. This revolution, therefore, might be made of a sudden by the opening of some vast cavern in the interior part of the globe, which an universal deluge must inevitably succeed; or possibly this change was not effected at once, but required a length of time, which I am rather inclined to think; however these conjectures may be, it is certain the revolution has occurred, and in my opinion very naturally; for to judge of the future, as well as the past, we must carefully attend to what daily happens before our eyes. It is a fact clearly established by repeated observations of travellers, that the ocean has a constant motion from the East to West; this motion, like the trade winds, is not only felt between the tropics, but also throughout the temperate climates, and as near the poles as navigators have gone; of course the Pacific Ocean makes a continual effort against the coasts of Tartary, China, and India; the Indian Ocean acts against the east coast of Africa; and the Atlantic in like manner against all the eastern coasts of America; therefore the sea must have always and still continues to gain land on the east and lose it on the west; and this alone is sufficient to prove the possibility of the change Of earth into sea, and sea into land. If, in fact, such are the effects of the sea's motion from east to west, may we not very reasonably suppose that Asia and the eastern continent is the oldest country in the world, and that Europe and part of Africa, especially the western coasts of these continents, as Great Britain, France, Spain, Muratania, &c. are of a more modern date? Both history and physics agree in confirming this conjecture.

There are, however, many other causes which concur with the continual motion of the sea from east to west, in producing these effects.

In many places there are lands lower than the level of the sea, and which are only defended from it by an isthmus of rocks, or by banks and dykes of still weaker materials; these barriers must gradually be destroyed by the constant action of the sea, when the lands will be overflowed, and constantly make part of the ocean. Besides, are not mountains daily decreasing by the rains, which loosen the earth, and carry it down into the vallies? It is also well known that floods wash the earth from the plains and high grounds into the small brooks and rivers, which in their turn convey it into the sea. By these means the bottom of the sea is filling up by degrees, the surface of the earth lowering to a level, and nothing but time is necessary for the sea's successively changing places with the earth.

I speak not here of those remote causes which stand above our comprehension;

of those convulsions of nature, whose least effects would be fatal to the world; the near approach of a comet, the absence of the moon, the introduction of a new planet, &c. are suppositions on which it is easy to give scope to the imagination. Such causes would produce any effects we chose, and from a single hypothesis of this nature, a thousand physical romances might be drawn, and which the authors might term, THE THEORY OF THE EARTH. As historians we reject these vain speculations; they are mere possibilities which suppose the destruction of the universe, in which our globe, like a particle of forsaken matter, escapes our observation, and is no longer an object worthy regard; but to preserve consistency, we must take the earth as it is, closely observing every part, and by inductions judge of the future from what exists at present; in other respects we ought not to be affected by causes which seldom happen, and whose effects are always sudden and violent; they do not occur in the common course of nature; but effects which are daily repeated, motions which succeed each other without interruption, and operations that are constant, ought alone to be the ground of our reasoning.

We will add some examples thereto; we will combine particular effects with general causes, and give a detail of facts which will render apparent, and explain the different changes that the earth has undergone, whether by the eruption of the sea upon the land, or by retiring from that which it had formerly covered.

The greatest eruption was certainly that which gave rise to the Mediterranean sea. The ocean flows through a narrow channel between two promontories with great rapidity, and then forms a vast sea, which, without including the Black sea, is about seven times larger than the kingdom of France. Its motion through the straits of Gibraltar is contrary to all other straits, for the general motion of the sea is from east to west, but in that alone it is from the west to the east, which proves that the Mediterranean sea is not an ancient gulph, but that it has been formed by an eruption, produced by some accidental cause; as an earthquake which might swallow up the earth in the strait, or by a violent effort of the ocean, caused by the wind, which might have forced its way through the banks between the promontories of Gibraltar and Ceuta. This opinion is authorised by the testimony of the ancients, who declare in their writings, that the Mediterranean sea did not formerly exist; and confirmed by natural history and observations made on the opposite coasts of Spain, where similar beds of stones and earth are found upon the same levels, in like manner as they are in two mountains, separated by a small valley.

The ocean having forced this passage, it ran at first through the straits with much greater rapidity than at present, and overflowed the continent that joined Europe to Africa. The waters covered all the low countries, of which we can only now perceive the tops of some of the considerable mountains, such as parts of Italy, the islands of Sicily, Malta, Corsica, Sardinia, Cyprus, Rhodes, and those of the Archipelago.

In this eruption I have not included the Black sea, because the quantity of water it receives from the Danube, Nieper, Don, and various other rivers, is fully sufficient to form and support it; and besides, it flows with great rapidity through the Bosphorus into the Mediterranean. It might also be presumed that the Black and Caspian seas were formerly only two large lakes, joined by a narrow communication, or by a morass, or small lake, which united the Don and the Wolga near Tria, where these two rivers flow near each other; nor is it improbable that these two seas or lakes were then of much greater extent, for the immense rivers which fall into the Black and Caspian seas may have brought down a sufficient quantity of earth to shut up the communication, and form that neck of land by which they are now separated; for we know great rivers, in the course of time, fill up seas and form new land, as the province at the mouth of the Yellow river in China; Louisania at the mouth of the Mississippi, and the northern part of Egypt, which owes its existence to the inundations of the Nile; the rapidity of which brings down such quantities of earth from the internal parts of Africa, as to deposit on the shores, during the inundations, a body of slime and mud of more than fifty feet in depth. The province of the Yellow river and Louisania have, in like manner, been formed by the soil from the rivers.

The Caspian sea is actually a real lake; having no communication with other seas, not even with the lake Aral, which seems to have been a part of it, being only separated from it by a large track of sand, in which neither rivers nor canals for communication the waters have as yet been found. This sea, therefore, has no external communication with any other; and I do not know that we are authorised to suspect that it has an internal one with the Black sea, or with the Gulph of Persia. It is true the Caspian sea receives the Wolga and many other rivers, which seem to furnish it with more water than is lost by evaporation; but independent of the difficulty of such calculation, if it had a communication with any other sea, a constant and rapid current towards the opening would have marked its course, and I never heard of any such discovery being made; travellers of the best credit affirm to the contrary, and consequently the Caspian sea must lose by evaporation just as much water as it receives from the Wolga and other rivers.

Nor is it any improbable conjecture that the Black sea will at some period be separated from the Mediterranean; and that the Bosphorus will be shut up, whenever the great rivers shall have accumulated a sufficient quantity of earth to answer that effect; this may be the case in the course of time by the successive diminution of waters in rivers, in proportion as the mountains from whence they draw their sources are lowered by the rains, and those other causes we have just alluded to.

The Caspian and Black seas must therefore be looked upon rather as lakes than gulphs of the ocean, for they resemble other lakes which receive a number of rivers without any apparent outlet, such as the Dead sea, many lakes in Africa and other places. These two seas are not near so salt as the Mediterranean or the ocean; and all voyagers affirm that the navigation in the Black and Caspian seas, upon account of its shallowness and quantity of rocks and quicksands, is so extremely dangerous, that only small vessels can be used with safety which farther proves they must not be looked upon as gulphs of the ocean, but as immense bodies of water collected from great rivers.

A considerable eruption of the sea would doubtless take place upon the earth, if the isthmus which separates Africa from Asia was divided, as the Kings of Egypt, and afterwards the Caliphs projected; and I do not know that the communication between the Red sea and Mediterranean is sufficiently established, as the former must be higher than the latter. The Red sea is a narrow branch of the ocean, and does not receive into it a single river on the side of Egypt, and very few on the opposite coast; it will not therefore be subject to diminution, like those seas and lakes which are constantly receiving slime and sand from those rivers that flow into them. The ocean supplies the Red sea with all its water, and the motion of the tides is very evident in it, of course it must be affected by every movement of the ocean. But the Mediterranean must be lower than the ocean, because the current passes with great rapidity through the straits; besides, it receives the Nile, which flows parallel to the west coast of the Red sea, and which divides Egypt, a very low country; from all which it appears probable, that the Red sea is higher than the Mediterranean, and that if the isthmus of Suez was cut through, there Would be a great inundation, and a considerable augmentation of the Mediterranean would ensue; at least if the waters were not restrained by dykes and sluices placed at proper distances, and which was most likely the case if the ancient canal of communication ever had existence.

Without dwelling longer upon conjectures, which, although well founded, may appear hazardous and rash, we shall give some recent and certain examples of the change of the sea into land, and the land into sea. At Venice the bottom of the Adriatic is daily rising, and if great care had not been taken to clean and empty the canals, the whole would long since have formed part of the continent; the same may be said of most ports, bays, and mouths of rivers. In Holland the bottom of the sea has risen in many places; the gulph of Zuyderzee and the strait of the Texel cannot receive such large vessels as formerly. At the mouth of all rivers we find small islands, and banks of sand and earth brought down by the waters, and it is certain the sea will be filled up in every part where great rivers empty themselves. The Rhine is lost in the sands which itself accumulated. The Danube and the Nile, and all great rivers, after bringing down much sand and earth, no longer come to the sea by a single channel; they divide into different branches, and the intervals are filled up by the materials they have themselves brought thither. Morasses daily dry up; lands forsaken by the sea are cultivated; we navigate countries now covered by waters; in short, we see so many instances of land changing into water, and water into land, that we must be convinced of these alterations having, and will continue to take place; so that in time gulphs will become continents; isthmusses, straits; morasses, dry lands; and the tops of our mountains, the shoals of the sea.

Since then the waters have covered, and may successively cover, every part of the present dry land, our surprise must cease at finding every where marine productions and compositions, which could only be the works of the waters. We have already explained how the horizontal strata of the earth were formed, but the perpendicular divisions that are commonly found in rocks, clays, and all matters of which the globe is composed, still remain to be considered. These perpendicular stratas are, in fact, placed much farther from each other than the horizontal, and the softer the matter the greater the distance; in marble and hard earths they are frequently found only a few feet; but if the mass of rock be very extensive, then these fissures are at some fathoms distant; sometimes they descend from the top of the rock to the bottom, and sometimes terminate at an horizontal fissure. They are always perpendicular in the strata of calcinable matters, as chalk, marle, marble, &c. but are more oblique and irregularly placed in vitrifiable substances, brown freestone, and rocks of flint, where they are frequently adorned with chrystals, and other minerals. In quarries of marble or calcinable stone, the divisions are filled with spar, gypsum, gravel, and an earthy sand, which contains a great quantity of chalk. In clay, marls, and every other kind of earth, excepting turf, these perpendicular divisions are either empty or filled with such matters as the water has transported thither.

We need seek very little farther for the cause and origin of those perpendicular cracks. The materials by which the different strata are composed being carried by the water, and deposited as a kind of sediment, must necessarily, at first, contain a considerable share of water, the which, as they began to harden, they would part with by degrees, and, as they must necessarily lessen in the course of drying, that decrease would occasion them to split at irregular distances. They naturally split in a perpendicular direction, because in that direction the action of gravity of one particle upon another has no actual effect, while, on the contrary, it is directly opposite in a horizontal situation; the diminution of bulk therefore could have no sensible effect but in a vertical line. I say it is the diminution of drying, and not the contained water forcing a place to issue, is the cause of these perpendicular fissures, for I have often observed that the two sides of those fissures answer throughout their whole height, as exactly as two sides of a split piece of wood; their insides are rough and irregular, whereas if they had been made by the motion of the water, they would have been smooth and polished; therefore these cracks must be produced suddenly and at once or by degrees in drying, like the flaws in wood, and the greatest part of the water they contained evaporated through the pores. The divisions of these perpendicular cracks vary greatly as to the extent of their openings; some of them being not more than half an inch, others increasing to one or two feet; there are some many fathoms, and which form those precipices so often met with in the Alps and other high mountains. The small ones are produced by drying alone, but those which extend to several feet are the effects of other causes; for instance, the sinking of the foundation on one side while the other remains unmoved; if the base sinks but a line or two, it is sufficient to produce openings of many feet in a rock of considerable height. Sometimes rocks, which are founded on clay or sand, incline to one side, by which motion the perpendicular cracks become extended.

I have not yet mentioned those large openings which are found in rocks and mountains; they must have been produced by great sinkings, as of immense caverns, unable longer to support the weight with which they were encumbered, but these intervals are very different from perpendicular fissures; they appear to be vacancies opened by the hand of Nature for the communication of nations. In this manner all vacancies in large mountains and divisions, by straits in the sea, seem to present themselves; such as the straits of Thermopylæ, the ports of Caucasus, the Cordeliers, the extremity of the straits of Gibraltar, the entrance of the Hellespont, &c. These could not have been occasioned by the simple separation by drying of matter, but by considerable parts of the lands themselves being sunk, swallowed up, or overturned.

These great sinkings, though produced by accidental causes, hold a first place in the principal circumstances in the History of the Earth, and not a little contributed to change the face of the Globe; the greatest part of them have been produced by subterraneous fires, whose explosions cause earthquakes and volcanos; the force of these inflamed and confined matters in the bowels of the earth is beyond compare; by it cities have been swallowed up, provinces overturned, and mountains overthrown. But however great this force may be, and prodigious as the effects appear, we cannot assent to the opinion of those authors who suppose these subterraneous fires proceed from an immense abyss of flame in the centre of the earth, neither give we credit to the common notion that they proceed from a great depth below the surface of the earth, air being absolutely necessary for the support of inflammation. In examining the materials which issue from volcanos, even in the most violent eruptions, it appears very plain, that the furnace of the inflamed matters is not at any great depth, as they are similar to those found in mountains, disfigured only by the calcination, and the melting of the metallic parts which they contain; and to be convinced that the matters cast out by volcanos do not come from any great depth, we have only to consider of the height of the mountain, and judge of the immense force that would be necessary to cast up stones and minerals to the height of half a league; for Ætna, Hecla, and many other volcanos have at least that elevation from the plains. Now it is perfectly well known that the action of fire is equal in every direction; it cannot therefore act upwards, with a force capable of throwing large stones half a league high, without an equal re-action downwards, and on the sides, and which re-action must very soon pierce and destroy the mountain on every side, because the materials which compose it are not more dense and firm than those thrown out; how then can it be imagined that the cavity, which must be considered as the type or cannon, could resist so great a force as would be necessary to raise those bodies to the mouth of the volcano? Besides, if this cavity was deeper, as the external orifice is not great, it would be impossible for so large a quantity of inflamed and liquid matter to issue out at once, without clashing against each other, and against the sides of the tube, and by passing through so long a space they would run the chance of being extinguished and hardened. We often see rivers of bitumen and melted sulphur, thrown out of the volcanos with stones and minerals, flow from the tops of the mountains into the plains; is it natural to imagine that matters so fluid, and so little able to resist violent action, should be elevated from any great depth? All the observations that can be made on this subject will prove that the fire of the volcano is not far from the summit of the mountain, and that it never descends to the level of the plain.

This idea of volcanos does not, however, render it inconsistent that they are the cause of earthquakes, and that their shocks may be felt on the plains to very considerable distances; nor that one volcano may not communicate with another by means of subterraneous passages; but it is of the depth of the fire's confinement that we now speak, and which can only be at a small distance from the mouth of the volcano. It is not necessary to produce an earthquake on a plain, that the bottom of the volcano should be below the level of that plain; nor that there should be internal cavities filled with the same combustible matter, for a violent explosion, such as generally attends an eruption, may, like that of a powder magazine give so great a shock by its re-action, as to produce an earthquake that might be felt at a considerable distance.

I do not mean to say that there are no earthquakes produced by subterraneous fires, but merely that there are some which proceed only from the explosion of volcanos. In confirmation of what has been advanced on this subject, it is certain that volcanos are seldom met with on plains; on the contrary, they are constantly found in the highest mountains, and their mouths at the very summit of them. If the internal fires of the volcanos extended below the plains, would not passages be opened in them during violent eruptions? In the first eruption would not these fires rather have pierced the plains, where, by comparison, the resistance must be infinitely weaker, than force their way through a mountain more than half a league in height.

The reason why volcanos appear alone in mountains, is, because much greater quantities of minerals, sulphur, and pyrites, are contained in mountains, and more exposed than in the plains; besides which, those high places are more subject to the impressions of air, and receive greater quantities of rain and damps, by which mineral substances are capable of being heated and fermented into an absolute state of inflammation.

In short, it has often been observed, that, after violent eruptions, the mountains have shrunk and diminished in proportion to the quantity of matter which has been thrown out; another proof that the volcanos are not situated at the bottom of the mountain, but rather at no great distance from the summit itself.

In many places earthquakes have formed considerable hollows, and even separations in mountains; all other inequalities have been produced at the same time with the mountains themselves by the currents of the sea, for in every place where there has not been a violent convulsion, the strata of the mountains are parallel, and their angles exactly correspond. Those subterraneous caverns which have been produced by volcanos are easily to be distinguished from those formed by water; for the water, having washed away the sand and clay with which they are filled, leaves only the stones and rocks, and this is the origin of caverns upon hills; while those found upon the plains are commonly nothing but ancient pits and quarries, such as the salt quarries of Maestricht, the mines of Poland, &c. But natural caverns belong to mountains: they receive the water from the summit and its environs, from whence it issues over the surface wherever it can obtain a passage; and these are the sources of springs and rivers, and whenever a cavern is filled by any part falling in, an inundation generally ensues.

From what we have related, it is easy to be seen how much subterraneous fires contribute to change the surface and internal part of the globe. This cause is sufficiently powerful to produce very great effects: but it is difficult to conceive how the winds should occasion any sensible alterations upon the earth. The sea appears to be their empire, and indeed, excepting the tides, nothing has so powerful an influence upon the ocean; even the flux and reflux move in an uniform manner, and their effects are regularly the same; but the action of the winds is capricious and violent; they sometimes rush on with such impetuosity, and agitate the sea with such violence, that from a calm, smooth, and tranquil plain, it becomes furrowed with waves rolling mountains high, and dashing themselves to pieces against the rocks and shores. The winds cause constant alterations on the surface of the sea, but the surface of the land, which has so solid an appearance, we should suppose would not be subject to similar effects; by experience, however, it is known that the winds raise mountains of sand in Arabia and Africa; and that they cover plains with it; they frequently transport sand to great distances, and many miles into the sea, where they accumulate in such quantities as to form banks, downs, and even islands. It is also known that hurricanes are the scourge of the Antilles, Madagascar, and other countries, where they act with such fury, as to sweep away trees, plants, and animals, together with the soil which gave them subsistence: they cause rivers to ascend and descend, and produce new ones; they overthrow rocks and mountains; they make holes and gulphs on the earth, and entirely change the face of those unfortunate countries where they exist. Happily there are but few climates exposed to the impetuosity of those dreadful agitations of the air.

But the greatest and most general changes in the surface of the earth are produced by rains, floods, and torrents from the high lands. Their origins proceed from the vapours which the sun raises above the surface of the ocean, and which the wind transports through every climate. These vapours, which are sustained in the air, and conveyed at the will of the winds, are stopped in their progress by the tops of the hills which they encounter, where they accumulate until they become clouds and fall in the form of rain, dew, or snow. These waters at first descend upon the plains without any fixed course, but by degrees hollow out a bed for themselves; by their natural bent they run to the bottom of mountains, and penetrating or dissolving the land easiest to divide, they carry earth and sand away with them, cut deep channels in the plains, form themselves into rivers, and open a passage into the sea, which constantly receives as much water from the land rivers as it loses by evaporation. The windings in the channels of rivers have sinuosities, whose angles are correspondent to each other, so that where the waves form a saliant angle on one side, the other has an exactly opposite one; and as hills and mountains, which may be considered as the banks of the vallies which separate them, have also sinuosities in corresponding angles, it seems to demonstrate that the vallies have been formed, by degrees, by the currents of the sea, in the same manner as the rivers have hollowed out their beds on the earth.

The waters which run on the surface of the earth, and support its verdure and fertility, are not perhaps one half of those which the vapours produce; for there are many veins of water which sink to great depths in the internal part of the earth. In some places we are certain to meet with water by digging; in others, not any can be found. In almost all vallies and low grounds water is certain to be met with at moderate depths; but, on the contrary, in all high places it cannot be extracted from the bowels of the earth, but must be collected from the heavens. There are countries of great extent where a spring cannot be found, and where all the water which supplies the inhabitants and animals with drink is contained in pools and cisterns. In the east, especially in Arabia, Egypt, and Persia, wells are extremely scarce, and the people have been obliged to make reservoirs of a considerable extent to collect the waters as it falls from the heavens. These works, projected and executed from public necessity, are the most beautiful and magnificent monuments of the eastern nations; some of the reservoirs occupy a space of two square miles, and serve to fertilize whole provinces, by means of baths and small rivulets that let it out on every side. But in low countries, where the greatest rivers flow, we cannot dig far from the surface, without meeting with water, and in fields situate in the environs of rivers it is often obtained by a few strokes with a pick-axe.

The water, found in such quantities in low grounds, comes principally from the neighbouring hills and eminences; at the time of great rains or sudden melting of snow, a part of the water flows on the surface, but most of it penetrates through the small cracks and crevices it finds in the earth and rocks. This water springs up again to the surface wherever it can find vent; but it often filters through the sand until it comes to a bottom of clay or solid earth, where it forms subterraneous lakes, rivulets, and perhaps rivers, whose courses are entirely unknown; they must, however, follow the general laws of nature, and constantly flow from the higher grounds to the lower, and consequently these subterraneous waters must, in the end, fall into the sea, or collect in some low place, either on the surface or in the interior part of the earth; for there are several lakes into which no rivers enter, nor from which there are not any issue; and a much greater number, which do not receive any considerable river, that are the sources of the greatest rivers on earth; such as the lake of St. Laurence; the lake Chiamè, from whence spring two great rivers that water the kingdoms of Asam and Pegu; the lake of Assiniboil in America; those of Ozera in Muscovy, that give rise to the river Irtis, and a great number of others. These lakes, it is evident, must be produced by the waters from the high lands passing through subterraneous passages, and collecting in the lowest places. Some indeed have asserted that lakes are to be found on the summit of the highest mountains; but to this no credit can be given, for those found on the Alps, and other elevated places, are all surrounded by much more lofty mountains, and derive their origin from the waters which run down the sides, or are filtered through those eminences in the same manner as the lakes in the plains obtain their sources from the neighbouring hills which overtop them.

It is apparent, therefore, that lakes have existence in the bowels of the earth, especially under large plains and extensive vallies. Mountains, hills, and all eminences have either a perpendicular or inclined situation, and are exposed on all sides; the waters which fall on their summits, after having penetrated into the earth, cannot fail, from the declivity of the ground, of finding issue in many places, and breaking in forms out of springs and fountains, and consequently there will be little, if any water, remain in the mountains. On the contrary, in plains, as the water which filters through the earth can find no vent, it must collect in subterraneous caverns, or be dispersed and divided among sand and gravel. It is these waters which are so universally diffused through low grounds. The bottom of a pit or well is nothing else but a kind of bason into which the waters that issue from the adjoining lands insinuate themselves, at first falling drop by drop, but afterwards, as the passages are opened, it receives supplies from greater distances, and then continually runs in little streams or rills; from which circumstance, although we can find water in any part of a plain, yet we can obtain a supply but for a certain number of wells, proportionate to the quantity of water dispersed, or rather to the extent of the higher lands from whence they come.

It is unnecessary to dig below the level of the river to find water; it is generally met with at much less depths, and there is no appearance that waters of rivers filter far through the earth. The origin of waters found in the earth below the level of rivers is not to be attributed to them; for in rivers or torrents which are dried up, or whose courses have been turned, we find no greater quantity of water by digging in their beds than in the neighbouring lands at an equal depth.

A piece of land of five or six feet in thickness is sufficient to contain water, and prevent it from escaping; and I have often observed that the banks of brooks and pools are not sensibly wet at six inches distance from the water.

It is true that the extent of the filtration is in proportion as the soil is more or less penetrable; but if we examine the standing pools with sandy bottoms, we shall perceive the water confined in the small compass it had hollowed itself, and the moisture spread but a very few inches; even in vegetable earth it has no great extent, which must be more porous than sand or hard soil. It is a certain fact, that in a garden we may almost inundate one bed without those nearly adjoining feeling any moisture from it[2]. I have examined pieces of garden ground, eight or ten feet thick, which had not been stirred for many years, and whose surface was nearly level, and found that the rain water never penetrated deeper than three or four feet; and on turning it up in the spring, after a wet winter, I found it as dry as when first heaped together.

I made the same observation on earth which had laid in ridges two hundred years; below three or four feet it was as dry as dust; from which it is plain that water does not extend so far by filtration as has been generally imagined.

By this means, therefore, the internal part of the earth can be supplied with a very small part; but water by its own weight descends from the surface to the greatest depths; it sinks through natural conduits, or penetrates small passages for itself; it follows the roots of trees, the cracks in rocks, the interstices in the earth, and divides and extends on all sides into an infinity of small branches and rills, always descending until its passage is opposed by clay or some solid body, where it continues collecting, and at length breaks out in form of springs upon the surface.

It would be very difficult to make an exact calculation of the quantity of subterraneous waters which have no apparent vent. Many have pretended that it greatly surpasses all the waters that are on the surface of the earth.

Without mentioning those who have advanced that the interior part of the globe is entirely filled with water, there are some who believe there are an infinity

[2] These facts are so easily demonstrated, that the smallest observation will prove their veracity.

of floods, rivulets, and lakes in the bowels of the earth. But this opinion does not seem to be properly founded, and it is more probable that the quantity of subterraneous water, which never appears on the surface, is not very considerable; for if these subterraneous rivers are so very numerous, why do we never see any of their mouths forcing their way through the surface? Besides, rivers, and all running waters, produce great alterations on the surface of the earth; they transport the soil, wear away the most solid rocks, and displace all matters which oppose their passage. It would certainly be the same in subterraneous rivers; the same effects would be produced; but no such alterations have ever as yet been observed; the different strata remains parallel, and every where preserves its original position; and it is but in a very few places that any considerable subterraneous veins of water have been discovered. Thus water in the internal part of the earth, though great, acts but in a small degree, as it is divided in an infinity of little streams, and retained by a number of obstacles; and being so generally dispersed, it gives rise to many substances totally different from primitive matters, both in form and organization.

From all these observations we may fairly conclude, that it is the continual motion of the flux and reflux of the sea which has produced mountains, vallies, and other inequalities on the surface of the earth; that it is the currents of the ocean which have hollowed vallies, raised hills, and given them corresponding directions; that it is those waters of the sea which, by transporting earth, &c. and depositing them in horizontal layers, have formed the parallel strata; that it is the waters from heaven, which by degrees destroy the effects of the sea, by continually lowering the summit of mountains, filling up vallies, and stopping the mouths of gulphs and rivers, and which, by bringing all to a level, will, in the course of time, return this earth to the sea, which, by its natural operations, will again form new continents, containing vallies and mountains exactly similar to those which we at present inhabit.

PROOF OF THE THEORY OF THE EARTH.

ARTICLE I.

ON THE FORMATION OF THE PLANETS.

Our subject being Natural History, we would willingly dispense with astronomical observations; but as the nature of the earth is so closely connected with the heavenly bodies, and such observations being calculated to illustrate more fully what has been said, it is necessary to give some general ideas of the formation, motion, figure of the earth and other planets.

The earth is a globe of about three thousand leagues diameter; it is situate one thousand millions of leagues from the sun, around which it makes its revolution in three hundred and sixty-five days. This revolution is the result of two forces; the one may be considered as an impulse from right to left, or from left to right, and the other an attraction from above downwards, or beneath upwards, to a common centre. The direction of these two forces, and their quantities, are so nicely combined and proportioned, that they produce an almost uniform motion in an ellipse, very near to a circle. Like the other planets the earth is opaque, it throws out a shadow; it receives and reflects the light of the sun, round which it revolves in a space of time proportioned to its relative distance and density. It also turns round its own axis once in twenty-four hours, and its axis is inclined 66-1/4 degrees on the plane of the orbit. Its figure is spheroidical, the two axes of which differ about 160th part from each other, and the smallest axis is that round which the revolution is made.

These are the principal phenomena of the earth, the result of discoveries made by means of geometry, astronomy, and navigation. We shall not here enter into the detail of the proofs and observations by which those facts have been ascertained, but only make a few remarks to clear up what is still doubtful, and at the same time give our ideas respecting the formation of the planets, and the different changes thro' which it is possible they have passed before they arrived at the state we at present see them.

There have been so many systems and hypotheses framed upon the formation of the terrestrial globe, and the changes which it has undergone, that we may presume to add our conjectures to those who have written upon the subject, especially as we mean to support them with a greater degree of probability than has hitherto been done: and we are the more inclined to deliver our opinion upon this subject, from the hope that we shall enable the reader to pronounce on the difference between an hypothesis drawn from possibilities, and a theory founded in facts; between a system, such as we are here about to present, on the formation and original state of the earth, and a physical history of its real condition, which has been given in the preceding discourse.

Galileo having found the laws of falling bodies, and Kepler having observed that the area described by the principal planets in moving round the sun, and those of the satellites round the planets to which they belong, are proportionable to the time of their revolutions, and that such periods were also in proportion to the square roots of the cubes of their distances from the sun, or principal planets. Newton found that the force which caused heavy bodies to fall on the surface of the earth, extended to the moon, and retained it in its orbit; that this force diminished in the same proportion as the square of the distance increased, and consequently that the moon is attracted by the earth; that the earth and planets are attracted by the sun; and that in general all bodies which revolve round a centre, and describe areas proportioned to the times of their revolution, are attracted towards that point. This power, known by the name of GRAVITY, is therefore diffused throughout all matter; planets, comets, the sun, the earth, and all nature, is subject to its laws, and it serves as a basis to the general harmony which reigns in the universe. Nothing is better proved in physics than the actual existence of this power in every material substance. Observation has confirmed the effects of this power, and geometrical calculations have determined the quantity and relations of it.

This general cause being known, the effects would easily be deduced from it, if the action of the powers which produce it were not too complicated. A single moment's reflection upon the solar system will fully demonstrate the difficulties that have attended this subject; the principal planets are attracted by the sun, and the sun by the planets; the satellites are also attracted by their principal planets, and each planet attracts all the rest, and is attracted by them. All these actions and reactions vary according to the quantities of matter and the distances, and produce great inequalities and irregularities. How is so great a number of connections to be combined and estimated? It appears almost impossible in such a crowd of objects to follow any particular one; nevertheless those difficulties have been surmounted, and calculation has confirmed the suppositions of them, each observation is become a new demonstration, and the systematic order of the universe is laid open to the eyes of all those who can distinguish truth from error.

We feel some little stop, by the force of impulsion remaining unknown; but this, however, does not by any means affect the general theory. We evidently see the force of attraction always draws the planets towards the sun, they would fall in a perpendicular line, on that planet, if they were not repelled by some other power that obliges them to move in a straight line, and which impulsive force would compel them to fly off the tangents of their respective orbits, if the force of attraction ceased one moment. The force of impulsion was certainly communicated to the planets by the hand of the Almighty, when he gave motion to the universe; but we ought as much as possible to abstain in physics from having recourse to supernatural causes; and it appears that a probable reason may be given for this impulsive force, perfectly accordant with the law of mechanics, and not by any means more astonishing than the changes and revolutions which may and must happen in the universe.

The sphere of the sun's attraction does not confine itself to the orbs of the planets, but extends to a remote distance, always decreasing in the same ratio as the square of the distance increases; it is demonstrated that the comets which are lost to our sight, in the regions of the sky, obey this power, and by it their motions, like that of the planets, are regulated. All these stars, whose tracts are so different, move round the sun, and describe areas proportioned to the time; the planets in ellipses more or less approaching a circle, and the comets in narrow ellipses of a great extent. Comets and planets move, therefore, by virtue of the force of attraction and impulsion, which continually acting at one time obliges them to describe these courses; but it must be remarked that comets pass over the solar system in all directions, and that the inclinations of their orbits are very different, insomuch that, although subject like the planets to the force of attraction, they have nothing in common with respect to their progressive or impulsive motions, but appear in this respect independent of each other: the planets, on the contrary, move round the sun in the same direction, and almost in the same plane, never exceeding 7-1/2 degrees of inclination in their planes, the most distant from their orbits. This conformity of position and direction in the motion of the planets, necessarily implies that their impulsive force has been communicated to them by one and the same cause.

May it not be imagined, with some degree of probability, that a comet falling into the body of the sun, will displace and separate some parts from the surface, and communicate to them a motion of impulsion, insomuch that the planets may formerly have belonged to the body of the sun, and been detached therefrom by an impulsive force, and which they still preserve.

This supposition appears to be at least as well founded as the opinion of Leibnitz, who supposes that the earth and planets had formerly been suns; and his system, of which an account will be given in the fifth article, would have been more comprehensive and more agreeable to probability, if he had raised himself to this idea. We agree with him in thinking that this effect was produced at the time when Moses said that God divided light from darkness; for, according to Leibnitz, light was divided from darkness when the planets were extinguished; but in our supposition there was a real physical separation, since the opaque bodies of the planets were divided from the luminous matter which composes the sun.

This idea of the cause of the impulsive force of the planets will be found much less objectionable, when an estimation is made of the analogies and degrees of probability, by which it may be supported. In the first place, the motion of the planets are in the same direction, from West to East, and therefore, according to calculation, it is sixty-four to one that such would not have been the case, if they had not been indebted to the same cause for their impulsive forces.

This, probably, will be considerably augmented by the second analogy, viz. that the inclination of the planes of the orbits do not exceed 7-1/2 degrees; for, by comparing the spaces, we shall find there is twenty-four to one, that two planets are found in their most distant places at the same time, and consequently [5], or 7,692,624 to one, that all six would by chance be thus placed; or, what amounts to the same, there is a great degree of probability that the planets have been impressed with one common moving force, and which has given them this position. But what can have bestowed this common impulsive motion, but the force and direction of the bodies by which it was originally communicated? It may therefore be concluded, with great probability, that the planets received their impulsive motion by one single stroke. This likelihood, which is almost equivalent to a certainty, being established, I seek to know what moving bodies could produce this effect, and I find nothing but comets capable of communicating a motion to such vast bodies.

By examining the course of comets, we shall be easily persuaded, that it is almost necessary for some of them occasionally to fall into the sun. That of 1680 approached so near, that at its perihelium it was not more distant from the sun than a sixteenth part of its diameter, and if it returns, as there is every appearance it will, in 2255, it may then possibly fall into the sun; that must depend on the rencounters it will meet with in its road, and of the retardment it suffers in passing through the atmosphere of the sun[3].

We may, therefore, presume with the great Newton, that comets sometimes fall into the sun; but this fall may be made in different directions. If they fall perpendicular, or in a direction not very oblique, they will remain in the sun, and serve for food to the fire which that luminary consumes, and the motion of impulsion which they will have communicated to the sun, will produce no other effect than that of removing it more or less, according as the mass of the comet will be more or less considerable; but if the fall of the comet is in a very oblique direction, which will most frequently happen, then the comet will only graze the surface of the sun, or slightly furrow it; and in this case it may drive out some parts of matter to which it will communicate a common motion of impulsion, and these parts so forced out of the body of the sun, and even the comet itself, may then become planets, and turn round this luminary in the same direction, and in almost the same plane. We might perhaps calculate what quantity of matter, velocity, and direction a comet should have, to impel from the sun an equal quantity of matter to that which the six planets and their satellites contain; but it will be sufficient to observe here, that all the planets, with their satellites, do not make the 650th part of the mass of the sun,[4] because the density of the large planets, Saturn and Jupiter, is less than that of the sun; and although the earth be four times, and the moon near five times more dense than the sun, they are nevertheless but as atoms in comparison with his extensive body.

However inconsiderable the 650th part may be, yet it certainly at first appears to require a very powerful comet to separate even that much from the body of the sun; but if we reflect on the prodigious velocity of comets in their perihelion, a velocity so much the greater as they approach nearer the sun; if, besides, we pay attention to the density and solidity of the matter of which they must be composed, to suffer, without being destroyed, the inconceivable heat they endure; and consider the bright and solid light which shines through their dark and immense atmospheres, which surround, and must obscure them, it cannot be doubted that

---

[3] Vide Newton, 2d edit. page 525.
[4] Vid. Newton, page 405.

the comets are composed of extremely solid and dense matters, and that they contain a greater quantity of matter in a small compass; that consequently a comet of no extraordinary bulk may have sufficient weight and velocity to displace the sun, and give a projectile motion to a quantity of matter, equal to the 650th part of the mass of this luminary. This perfectly agrees with what is known concerning the density of planets, which always decreases as their distance from the sun is increased, they having less heat to support; so that Saturn is less dense than Jupiter, and Jupiter much less than the earth; therefore if the density of the planets be, as Newton asserts, proportionable to the quantity of heat which they have to support, Mercury will be seven times more dense than the earth, and twenty-eight times denser than the sun; and the comet of 1680 would be 28,000 times denser than the earth, or 112,000 times denser than the sun, and by supposing it as large as the earth, it would contain nearly an equal quantity of matter to the ninth part of the sun, or by giving it only the 100th part of the size of the earth, its mass would still be equal to the 900th part of the sun. From whence it is easy to conclude, that such a body, though it would be but a small comet, might separate and drive off from the sun a 900th or a 650th part, particularly if we attend to the immense velocity with which comets move when they pass in the vicinity of the sun.

Besides this, the conformity between the density of the matter of the planets, that of the sun deserves some attention. It is well known, that, both on and near the surface of the earth, there are some matters 14 or 1500 times denser than others. The densities of gold and air are nearly in this relation. But the internal parts of the earth and planets are composed of a more uniform matter, whose comparative density varies much less; and the conformity in the density of the planets and that of the sun is such, that of 650 parts which compose the whole of the matter of the planets, there are more than 640 of the same density as the matter of the sun, and only ten parts out of these 650 which are of a greater density, for Saturn and Jupiter are nearly of the same density as the sun, and the quantity of matter which these planets contain, is at least 64 times greater than that of the four inferior planets, Mars, the Earth, Venus, and Mercury. We must therefore admit, that the matter of which the planets are generally composed is nearly the same as that of the sun, and that consequently the one may have been separated from the other.

But it may be said, if the comet, by falling obliquely on the sun, drove off the matter which compose the planets, they, instead of describing circles of which the sun is the centre, would, on the contrary, at each revolution, have returned to the same point from whence they departed, as every projectile would which might be thrown off with sufficient force from the surface of the earth, to oblige it to turn perpetually: for it is easy to demonstrate that such, in that instance, would be the case, and therefore that the projection of the planets from the sun cannot be attributed to the impulsion of a comet.

To this I reply, that the matter which composes the planets did not come from the sun, in ready formed globes, but in the form of torrents, the motion of the anterior parts of which were accelerated by that of the posterior; and that the attraction of the anterior parts also accelerated the motion of the posterior, and that this acceleration produced by one or other of these causes, or perhaps by both, might be so great as to change the original direction of the motion occasioned by the impulse of the comet, from which cause a motion has resulted, such as we at present observe in the planets; especially when it is considered the sun is displaced from its station by the shock of the comet. An example will render this more reasonable; let us suppose, that from the top of a mountain a musket ball is discharged, and that the strength of the powder was sufficient to send it beyond the semi-diameter of the earth, it is certain that this ball would pass round the earth, and at each revolution return to the spot from whence it had been discharged: but, if instead of a musket-ball, we suppose a rocket had been discharged, wherein the action of the fire being durable, would greatly accelerate the motion of impulsion; this rocket, or rather the cartouch which contained it, would not return to the same place like the musket-ball, but would describe an orbit, whose perigee would be much farther distant from the earth, as the force of acceleration would be greater, and have changed the first direction.

Thus, provided there had been any acceleration in the motion of impulsion communicated to the torrent of matter by the fall of the comet, it is probable that the planets formed in this torrent, acquired the motion which we know they have in the circles and ellipsis of which the sun is the centre and focus.

The manner in which the great eruptions of volcanos are made, may afford us an idea of this acceleration of motion. It has been remarked that when Vesuvius begins to roar and eject the inflamed matter it contains, the first cloud has but a small degree of velocity, but which is soon accelerated by the impulse of the second; the second by the action of a third, and so on, until the heavy mass of bitumen, sulphur, cinders, melted metal, and huge stones, appear like massive clouds, and although they succeed each other nearly in the same directions, yet they greatly change that of the first, and drive it far beyond what it would have reached of itself.

In answer to this objection, it may be further observed, that the sun having been struck by the comet, received a degree of motion by the impulse, which displaced it from its former situation; and that although this motion of the sun is at present too little sensible for the notice of astronomers, nevertheless it may still exist, and the sun describe a curve round the centre of gravity of the whole system and if this is so, as I presume it is, we see perfectly that the planets, instead of returning near the sun at each revolution, will, on the contrary, have described orbits, the points of the perihelion of which will be as far distant from the sun, as it is itself from the place it originally occupied.

It may also be said, that if this acceleration of motion is made in the same direction, no change in the perihelion will be produced: but can it be thought that in a torrent, the particles of which succeed each other, there has been no change of direction; it is, on the contrary, very probable that a considerable change did take place, sufficient to cause the planets to move in the course they at present occupy.

It may be further urged, that if the sun had been displaced by the shock of a comet, it would move uniformly, and that hence this motion being common to the whole system, no alteration was necessary; but might not the sun before the shock have had a motion round the centre of the cometry system, to which primitive motion the stroke of the comet may have added or diminished? and would not that fully account for the actual motion of the planets?

If these suppositions are not admitted, may it not be presumed, that in the stroke of the comet against the sun, there was an elastic force which raised the torrent above the surface of the sun, instead of directly impelling it? which alone would be sufficient to remove the perihelion, and give the planets the motion they have retained. This supposition is not without probability, for the matter of the sun may possibly be very elastic, since light, the only part of it we are acquainted with, seems, by its effects, to be perfectly so. I own that I cannot say whether it is by the one or the other of these reasons, that the direction of the first motion of the impulse of the planets has changed, but they suffice to shew that such an alteration is not only possible but even probable, and that is sufficient for my purpose.

But, without dwelling any longer on the objections which might be made, I shall pursue the subject, and draw the fair conclusions on the proofs which analogies might furnish in favour of my hypothesis: let us, therefore, first see what might happen when these planets, and particularly the earth, received their impulsive motion, and in what state they were after having been separated from the sun. The comet having, by a single stroke, communicated a projectile motion to a quantity of matter equal to the 650th part of the sun's mass, the light particles would of course separate from the dense, and form, by their mutual attraction, globes of different densities: Saturn being composed of the most gross and light parts, would be the most remote from the sun: Jupiter being more dense than Saturn, would be less distant, and so on. The larger and least solid planets are the most remote, because they received an impulsive motion stronger than the smallest, and more dense: for, the force of impulsion communicating itself according to the surface, the same stroke would have moved the grosser and lighter parts of the matter of the sun with more velocity than the smallest and more weighty; a separation therefore will be made of the dense parts of different degrees, so that the density of the sun being equal to 100, that of Saturn will be equal to 67, that of Jupiter to 94-1/2, that of Mars to 200, that of the Earth to 400, that of Venus to 800, and that of Mercury to 2800. But the force of attraction not communicating like that of impulsion, according to the surface, but acting on the contrary on all parts of the mass, it will have checked the densest portions of matter; and it is for this reason that the densest planets are the nighest the sun, and turn round that planet with greater rapidity than the less dense planets, which are also the most remote.

Jupiter and Saturn, which are the largest and principal planets of the solar system, have retained the relation between their density and impulsive motions, in the most exact proportions; the density of Saturn is to that of Jupiter as 67 to 94-1/2 and their velocities are nearly as 88-2/3 to 120-1/72, or as 67 to 90-11/16; it is seldom that pure conjectures can draw such exact relations. It is true, that by following this relation between the velocity and density of planets, the density of the earth ought to be only as 206-7/18, and not 400, which is its real density; from hence it may be conceived, that our globe was formerly less dense than it is at present. With respect to the other planets, Mars, Venus, and Mercury, as their densities are known only by conjecture, we cannot be certain whether this circumstance will destroy or confirm our hypothesis. The opinion of Newton is, that density is so much the greater, as the heat to which the planet is exposed is the stronger; and it is on this idea that we have just said that Mars is one time less dense than the Earth, Venus one time, Mercury seven times, and the comet in 1680, 28,000 times denser than the earth: but this proportion between the density of the planets and the heat which they sustain, seems not well founded, when we consider Saturn and Jupiter, which are the principal objects; for, according to this relation between the density and heat, the density of Saturn would be about 4-7/18, and that of Jupiter as 14-17/22, instead of 67 and 94-1/2, a difference too great to be admitted, and must destroy the principles upon which it was founded. Thus, notwithstanding the confidence which the conjectures of Newton merit, I can but think that the density of the planets has more relation with their velocity than with the degree of heat to which they are exposed. This is only a final cause, and the other a physical relation, the preciseness of which is remarkable in Jupiter and Saturn: it is nevertheless true, that the density of the earth, instead of being 206-7/8, is found to be 400, and that consequently the terrestrial globe must be condensed in this ratio of 206-7/8 to 400.

But have not the condensations of the planets some relation with the quantity of the heat of the sun which they sustain? If so, Saturn, which is the most distant from that luminary, will have suffered little or no condensation; and Jupiter will be condensed from 90-11/16 to 94-1/2. Now the heat of the sun in Jupiter being to that of the sun upon the earth as 14-17/22 are to 400, the condensations ought to be in the same proportion. For instance, if Jupiter be condensed, as 90-11/16 to 94-1/2, and the earth had been placed in his orbit, it would have been condensed from 206-7/8 to 215-990/1451, but the earth being nearer the sun, and receiving a heat, whose relation to that which Jupiter receives is from 400 to 14-17/22, the quantity of condensation it would have experienced on the orbit of Jupiter by the proportion of 400 to 14-17/22, which gives nearly 234-1/3 for the quantity which the earth would be condensed. Its density was 206-7/8, by adding the quantity of its acquired condensation, we find 400-7/8 for its actual density, which nearly approaches the real density 400, determined to be so by the parallax of the moon. As to other planets, I do not here pretend to give exact proportions, but only approximations, to point out that their densities have a strong relation to their velocity in their respective orbits.

The comet, therefore, by its oblique fall upon the surface of the sun, having driven therefrom a quantity of matter equal to the 650th part of its whole mass; this matter, which must be considered in a liquid state, will at first have formed a torrent, the grosser and less dense parts of which will have been driven the farthest, and the smaller and more dense, having received only the like impulsion, will remain nearest its source; the force of the sun's attraction would inevitably act upon all the parts detached from him, and constrain them to circulate around his body, and at the same time the mutual attraction of the particles of matter would form themselves into globes at different distances from the sun, the nearest of which necessarily moving with greater rapidity in their orbits than those at a distance.

But another objection may be started, and it may be said, if the matter which composes the planets had been separated from the sun, they, like him, would have been burning and luminous bodies, not cold and opaque, for nothing resembles a globe of fire less than a globe of earth and water; and by comparison, the matter of the earth and planets is perfectly different from that of the sun?

To this it may be answered, that in the separation the matter changed its form, and the light or fire was extinguished by the stroke which caused this motion of impulsion. Besides, may it not be supposed that if the sun, or a burning star, moved with such velocity as the planet, that the fire would soon be extinguished; and that is the reason why all luminous stars are fixed, and that those stars which are called new, and which have probably changed places, are frequently extinguished and lost? This remark is somewhat confirmed by what has been observed in comets; they must burn to the centre when they pass to their perihelium: nevertheless they do not become luminous themselves, they only exhale burning vapours, of which they leave a considerable part behind them in their course.

I own, that in a medium where there is very little or no resistance, fire may subsist and suffer a very great motion without being extinguished: I also own, that what I have just said extends only to the stars which totally disappear, and not to those which have periodical returns, and appear and disappear alternately without changing place in the heavens. The phenomena of these stars has been explained in a very satisfactory manner by M. de Maupertuis, in his discourse on the figures of the planets. But the stars which appear and afterwards disappear entirely, must certainly have been extinguished, either by the velocity of their motion, or some other cause. We have not a single example of one luminous star revolving round another; and among the number of planets which compose our system, and which move round the sun with more or less rapidity, there is not one luminous of itself.

It may also be added, that fire cannot subsist so long in the small as in large masses, and that the planets must have burnt for some time after they were separated from the sun, but were at length extinguished for want of combustible matter, as probably would be the sun itself, and for the same reason; but in a length of time as far beyond that which extinguished the planets, as it exceeds in quantity of matter. Be this as it may, the matter of which the planets are formed being separated from the sun, by the stroke of a comet, that appears a sufficient reason for the extinction of their fires.

The earth and planets at the time of their quitting the sun, were in a state of total liquid fire; in this state they remained only as long as the violence of the heat which had produced it; and which heat necessarily underwent a gradual decay: it was in this state of fluidity that they took their circular forms, and that their regular motions raised the parts of their equators, and lowered their poles. This figure, which agrees so perfectly with the laws of hydrostatics, I am of opinion with Leibnitz, necessarily supposes that the earth and planets have been in a state of fluidity, caused by fire, and that the internal part of the earth must be a vitrifiable matter, of which sand, granite, &c. are the fragments and scoria.

It may, therefore, with some probability, be thought that the planets appertained to the sun, that they were separated by a single stroke, which gave to them a motion of impulsion, and that their position at different distances from the sun proceeds only from their different densities. It now only remains, to complete this theory, to explain the diurnal motion of the planets, and the formation or the satellites; but this, far from adding difficulties to my hypothesis, seems, on the contrary, to confirm it.

For the diurnal motion, or rotation, depends solely on the obliquity of the stroke, an oblique impulse therefore on the surface of a body will necessarily give it a rotative motion; this motion will be equal and always the same, if the body which receives it is homogeneous, and it will be unequal if the body is composed of heterogeneous parts, or of different densities; hence we may conclude that in all the planets the matter is homogeneous, since their diurnal motions are equal, and regularly performed in the same period of time. Another proof that the separation of the dense or less dense parts were originally from the sun.

But the obliquity of the stroke might be such, as to separate from the body of the principal planet a small part of matter, which would of course continue to move in the same direction; these parts would be united, according to their densities, at different distances from the planet, by the force of their mutual attraction, and at the same time follow its course round the sun, by revolving about the body of the planet, nearly in the plane of its orbit. It is plain, that those small parts so separated are the satellites: thus the formation, position, and direction of the motions of the satellites perfectly agree with our theory; for they have all the same motion in concentrical circles round their principal planet; their motion is in the same direction, and that nearly in the plane of their orbits. All these effects, which are common to them, and which depend on an impulsive force, can proceed only from one common cause, which is, impulsive motion, communicated to them by one and the same oblique stroke.

What we have just said on the cause of the motion and formation of the satellites, will acquire more probability, if we consider all the circumstances of the phenomena. The planets which turn the swiftest on their axis, are those which have satellites. The earth turns quicker than Mars in the relation of about 24 to 15; the earth has a satellite, but Mars has none. Jupiter, whose rapidity round its axis is five to six hundred times greater than that of the earth, has four satellites, and there is a great appearance that Saturn, which has five, and a ring, turns still more quickly than Jupiter.

It may even be conjectured with some foundation, that the ring of Saturn is parallel to the equator of the planet, so that the plane of the equator of the ring, and that of Saturn, are nearly the same; for by supposing, according to the preceding theory, that the obliquity of the stroke by which Saturn has been set in motion was very great, the velocity around the axis will, at first, have been in proportion as the centrifugal force exceeds that of gravity, and there will be detached from its equator and neighbouring parts, a considerable quantity of matter, which will necessarily have taken the figure of a ring, whose plane must be nearly the same as that of the equator of the planet; and this quantity of matter having been detached from the vicinity of the equator of Saturn, must have lowered the equator of that planet, which causes that, notwithstanding its rapidity, the diameters of Saturn cannot be so unequal as those of Jupiter, which differ from each other more than an eleventh part.

However great the probability of what I have advanced on the formation of the planets and their satellites may appear to me, yet, every man has his particular measurement, to estimate probabilities of this nature; and as this measurement depends on the strength of the understanding to combine more or less distant relations, I do not pretend to convince the incredulous. I have not only thought it my duty to offer these ideas, because they appear to me reasonable, and calculated to clear up a subject, on which, however important, nothing has hitherto been written, but because the impulsive motion in the planets enter at least as one half of the composition of the universe, which gravity alone cannot unfold. I shall only add the following questions to those who are inclined to deny the possibility of my system.

1. Is it not natural to imagine, that a body in motion has received that motion by the stroke of another body?

2. Is it not very probable, that when many bodies move in the same direction, that they have received this direction by one single stroke, or by many strokes directed in the same manner?

3. Is it not more probable that when many bodies have the same direction in their motion, and are placed in the same plane, that they received this direction and this position by one and the same stroke, rather than by a number?

4. At the time a body is put in motion by the force of impulsion, is it not probable that it receives it obliquely, and, consequently, is obliged to turn on its axis so much the quicker, as the obliquity of the stroke will have been greater? If these questions should not appear unreasonable, the theory, of which we have presented the outlines, will cease to appear an absurdity.

Let us now pass on to something which more nearly concerns us, and examine the figure of the earth, on which so many researches and such great observations have been made. The earth being, as it appears by the equality of its diurnal motion and the constancy of the inclination of its axis, composed of homogeneous parts, which attract each other in proportion to their quantity of matter, it would necessarily have taken the figure of a globe perfectly spherical, if the motion of impulsation had been given it in a perpendicular direction to the surface; but this stroke having been obliquely given, the earth turned on its axis at the moment it took its form; and from the combination of this impulsive force, the attraction of the parts, there has resulted a spheroid figure, more elevated under the great circle of rotation, and lower at the two extremities of the axis, and this because the action of the centrifugal force proceeding from the diurnal rotation must diminish the action of gravity. Thus, the earth being homogeneous, and having received a rotative motion, necessarily took a spheroidical figure, the two axes of which differ a 230th part from each other. This may be clearly demonstrated, and does not depend on any hypothesis whatever. The laws of gravity are perfectly known, and we cannot doubt that bodies attract each other in a direct ratio of their masses, and in an inverted ratio, at the squares of their distances; so likewise we cannot doubt, that the general action of any body is not composed of all the particular actions of its parts. Thus each part of matter mutually attracts in a direct ratio of its mass and an inverted ratio of its distance, and from all these attractions there results a sphere when there is no rotatory motion, and a spheroid when there is one. This spheroid is longer or shorter at the two extremities of the axis of rotation, in proportion to the velocity of its diurnal motion, and the earth has then, by virtue of its rotative velocity, and of the mutual attraction of all its parts, the figure of a spheroid, the two axes of which are as 229 to 230 to one another.

Thus, by its original constituent, by its homogeneousness, and independent of every hypothesis from the direction of gravity, the earth has taken this figure of a spheroid at its formation, and agreeable to mechanical laws: its equatorial diameter was raised about 6-1/2 leagues higher than under the poles.

I shall dwell on this article, because there are still geometricians who think that the figure of the earth depends upon theory, and this from a system of philosophy they have embraced, and from a supposed direction of gravity. The first thing we have to demonstrate is, the mutual attraction of every part of matter, and the second the homogeneousness of the terrestrial globe; if we clearly prove, that these two circumstances are really so, there will no longer be any hypothesis to be made on the direction of gravity: the earth will necessarily have the figure Newton decided in favour of, and every other figure given to it by virtue of vortexes or other hypotheses, will not be able to subsist.

It cannot be doubted, that it is the force of gravity which retains the planets in their orbits; the satellites of Saturn gravitate towards Saturn, those of Jupiter towards Jupiter, the Moon gravitates towards the Earth: and Saturn, Jupiter, Mars, the Earth, Venus, and Mercury, gravitate towards the Sun: so likewise Saturn and Jupiter gravitate towards their satellites, the Earth gravitates towards the Moon, and the Sun towards the whole of the planets. Gravitation is therefore general and mutual in all the planetary system, for action cannot be exercised without a re-action; all the planets, therefore, act mutually one on the other. This mutual attraction serves as a foundation to the laws of their motion, and is demonstrated to exist by its effects. When Saturn and Jupiter are in conjunction, they act one on the other, and this attraction produces an irregularity in their motion round the Sun. It is the same with the Earth and the Moon, they also mutually attract each other; but the irregularities of the motion of the Moon, proceeds from the attraction of the Sun, so that the Earth, the Sun, and the Moon, mutually act one on the other. Now this mutual attraction of the planets, when the distances are equal, is proportional to their quantity of matter, and the same force of gravity which causes heavy matter to fall on the surface of the Earth, and which extends to the Moon, is also proportional to the quantity of matter; therefore the total gravity of a planet is composed of the gravity of each of its parts; from whence all the parts of the matter, either in the Earth or in the planets, mutually attract each other and the Earth, by its rotation round its own axis, has necessarily taken the figure of a spheroid, the axes of which are as 229 to 230. The direction of the weight must be perpendicular to the Earth's surface; consequently no hypothesis, drawn from the direction of gravity, can be sustained, unless the general attraction of the parts of matter be denied; but the existence of this mutual attraction is demonstrated by observations, and the experiment of pendulums prove, that its extension is general; therefore we cannot support an hypothesis on the direction of gravity without going against experience and reason.

Let us now proceed to examine whether the matter of which the terrestrial globe is composed be homogeneous. I admit, that if it is supposed the globe is more dense in some parts than in others, the direction of gravity must be different from what we have just assigned, and that the figure of the Earth would also differ agreeable to those suppositions. But what reason have we to make these suppositions? Why, for example, should we suppose that the parts near the centre are denser than those which are more remote? Are not all the particles which compose the globe collected together by their mutual attraction? hence, each particle is a centre, and there is no reason to believe, that the parts which surround the centre are denser than those which are about any other point. Besides, if one considerable part of the globe was denser than another, the axis of rotation would be found near the dense parts, and an inequality would ensue in the diurnal revolution; we should remark an inequality in the apparent motion of the fixed stars; they would appear to move more quick or slow in the zenith, or horizon, according as we should be placed on the denser or lighter parts of the earth; and the axis of the globe no longer passing through the centre of gravity, would also very sensibly change its position: but nothing like this ever happens; on the contrary, the diurnal motion of the earth is equal and uniform. At all parts of the Earth's surface, the stars appear to move with the same velocity at all heights, and if there be any rotation in its axis, it is so trifling as to have escaped observation: it must therefore be concluded, that the globe is homogeneous, or nearly so in all its parts.

If the earth was a hollow and void globe, and the crust of which, for example, not more than two or three miles thick; it would produce these effects. 1. The mountains would be such considerable parts of the whole thickness of the crust, that great irregularities in the motions of the Earth would be occasioned by the attraction of the Moon and Sun: for when the highest parts of the globe, as the Cordeliers, should have the Moon at noon, the attraction would be much stronger on the whole globe than when she was in the meridian of the lowest parts. 2. The attraction of mountains would be much more considerable than it is in comparison with the attraction of the whole globe, and experiments made at the mountain of Chimboraco, in Peru, would in this case give more degrees than they have given seconds for the deviation of the plumb line. 3. The weight of bodies would be greater on the tops of high mountains than on the planes; so that we should feel ourselves considerably heavier, and should walk with more difficulty in high than in low places. These observations, with many others that might be added, must convince us, that the inner parts of the globe is not void, but filled with a dense matter.

On the other hand, if below the depth of two or three miles, the earth was filled with a matter much more dense than any known, it would necessarily occur, that every time we descended to moderate depths, we should weigh much more, and the motion of pendulums would be more accelerated than in fact they are when carried from an eminence into a plain: thus, we may presume that the internal part of the Earth is filled with a matter nearly similar to that which composes its surface. What may complete our determination in favour of this opinion is, that in the first formation of the globe, when it took its present spheroidical figure, the matter which composed it was in fusion, and, consequently, all its parts were homogeneous, and nearly equally dense. From that time the matter on the surface, although originally the same with the interior, has undergone a variety of changes by external causes, which has produced materials of such different densities; but it must be remarked, that the densest matters, as gold and metals, are also those the most seldom to be met with, and consequently the greatest part of the matter at the surface of the globe has not undergone any very great changes with relation to its density; the most common materials, as sand and clay, differ very little, insomuch, that we may conjecture, with great probability, that the internal part of the earth is composed of a vitrified matter, the density of which is nearly the same as that of sand, and that consequently the terrestrial globe in general may be regarded as homogeneous.

Notwithstanding this, it may be urged, that although the globe was composed of concentrical strata of different densities, the diurnal motion might be equally certain, and the uniform inclination of the axis as constant and undisturbed as it could be, on the supposition of its being composed of homogeneous matter. I acknowledge it, but I ask at the same time, if there is any reason to believe that strata of different densities do exist? If these conclusions be not rather a desire to adjust the works of Nature to our own ideas? And whether in physics we ought to admit suppositions which are not founded on observations or analogy?

It appears, therefore, that the earth, by virtue of the mutual attraction of its parts and its diurnal motion, assumed the figure of a spheroid; that it necessarily took that form from being in a state of fluidity; that, agreeable to the laws of gravity and of a centrifugal force, it could have no other figure: that in the moment of its formation as at present, there was a difference between the two diameters equal to a 230th part, and that, consequently, every hypothesis in which we find greater or less difference are fictions which merit no attention.

But it may be said, if this theory is true, and if 229 to 230 is the just relation of the axis, why did the mathematicians, sent to Lapland and Peru, agree to the relation of 174 to 175? From whence does this difference arise between theory and practice? And is it not more reasonable to give the preference to practice and measures, especially when we have been taken by the most able mathematicians of Europe[5], and with all necessary apparatus to establish the result.

To this I answer, that I have paid attention to the observations made at the equator and near the polar circle; that I have no doubt of their being exact, and that the earth may possibly be elevated an 175th part more at the equator than at the poles. But, at the same time, I maintain my theory, and I see clearly how the two conclusions may be reconciled. This difference is about four leagues in the two

[5] M. de Maupertuis' Figure of the Earth.

axes, so that the parts at the equator are raised two leagues more than they ought to be, according to my theory; this height answers exactly to the greatest inequalities on the surface of the globe, produced by the motion of the sea, and the action of the fluids. I will explain; it appears that when the earth was formed, it must necessarily have taken, by virtue of the mutual attraction of its parts, and the action of the centrifugal force, a spheroidical figure, the axes of which differ a 230th part: the original earth must have had this figure, which it took when it was fluid, or rather liquified by the fire; but after its formation the vapours which were extended and rarefied, as in the atmosphere and tail of a comet, became condensed, and fell on the surface in form of air and water: and when these waters became agitated by the flux and reflux, the matters were, by degrees, carried from the poles towards the equatorial parts; so that the poles were lowered about a league, and those of the equator raised in the same proportion; this was not suddenly done, but by degrees in succession of time; the earth being also exposed to the action of the winds, air, and sun; all these irregular causes concurred with the flux and reflux to furrow its surface, hollow it into valleys, and raise it into mountains; and producing other inequalities and irregularities, of which, nevertheless, the greatest thickness does not exceed one league at the equator; this inequality of two leagues, is, perhaps, the greatest which can be on the surface of the earth, for the highest mountains are scarce above one league in height, and there is much probability of the sea's not being more at its greatest depth. The theory is therefore true, and practice may be so likewise; the earth at first could not be raised above 6-1/2 leagues more at the equator than the poles, but the changes which have happened to its surface might afterwards raise it still more. Natural History wonderfully confirms this opinion, for we have proved in the preceding discourse that the flux and reflux, and other motions of the water, have produced mountains and all the inequalities on the surface of the globe, that this surface has undergone considerable changes, and that at the greatest depths, as well as on the greatest heights, bones, shells and other wrecks of animals, which inhabit the sea and earth, are met with.

It may be conjectured, from what has been said, that to find ancient earth, and matters which have never been removed from the spot in which they were first placed, we must dig near the poles, where the bed of the earth must be thinner than in the Southern climates.

On the whole, if we strictly examine the measures by which the figure of the earth is determined, we shall perceive this hypothesis enters into such determination; for it supposes the earth to have the figure of a regular curve, whereas from the constant changes the earth is continually undergoing from a variety and combination of causes, it is almost impossible that it should have retained any regular figure, and hence the poles might, originally, only be flattened a 230th part, as Newton says, and as my theory requires. Besides, although we had exactly the length of the degree at the polar circle and equator, have we not also the length of the degree as exactly in France? And the measure of M. Picard, has it not been verified? Add to this that the augmentation and diminution in the motion of the pendulum, do not agree with the result drawn from measurement, and that, on the contrary, they differ very little from the theory of Newton. This is surely more than is requisite to convince us that the poles are not flattened more than a 230th part, and that if there is any difference, it can proceed only from the inequalities, which the water and other external causes have produced on its surface; but these inequalities being more irregular than regular, we must not form any hypothesis thereon, nor suppose, that the meridians are ellipses, or any other regular curves. From whence we perceive, that if we should successively measure many degrees of the earth in all directions, we still should not be certain by that alone, of the exact situation of the poles, nor whether they were depressed more or less than the 230th part.

May it not also be conjectured, that if the inclination of the axis of the earth has changed, it can only be produced by the changes which have happened to the surface, since all the rest of the globe is homogeneous; that consequently this variation is too little sensible to be perceived by astronomers, and that if the earth is not encountered with a comet, or deranged, by any other external cause, its axis will remain perpetually inclined as it is at present, and as it has always been?

In order not to omit any conjecture which appears reasonable, may it not be said, that as the mountains and inequalities which are on the surface of the earth have been formed by the flux and reflux of the sea, the mountains and inequalities which we remark on the surface of the moon, have been produced by a similar cause? they certainly are much higher than those of the earth, but then her tides are also much stronger, occasioned by the earth's being considerably larger than the moon, and consequently producing her tides with a superior force; and this effect would be much greater if the moon had, like the earth, a rapid rotation; but as the moon presents always the same surface to the earth, the tides cannot operate but in proportion to the motion arising from her libration, by which it alternatively discovers to us a segment of its other hemisphere; this, however, must produce a kind of flux and reflux, quite different from that of our sea, and the effects of which will be much less considerable than if the moon had from its course a revolution round its axis, as quick as the rotation of the terrestrial globe.

I should furnish a volume as large as that of Burnet or Whiston's, if I were to enlarge on the ideas which arise in support of the above; by giving them a geometrical air, in imitation of the last author, I might add considerably to their weight; but, in my opinion, hypothesis, however probable, ought not to be treated with such pomposity; it being a dress which borders so much on quackery.

ARTICLE II.

FROM THE SYSTEM OF WHISTON[6].

This Author commences his treatise by a dissertation on the creation of the world; he says that the account of it given by Moses in the text of Genesis has not been rightly understood; that the translators have confined themselves too much to the letter and superficial views, without attending to nature, reason, and philosophy. The common notion of the world being made in six days, he says is absolutely false, and that the description given by Moses, is not an exact and philosophical narration of the creation and origin of the universe, but only an historical representation of the terrestrial globe. The earth, according to him, existed in the chaos; and, at the time mentioned by Moses, received the form, situation and consistency necessary to be inhabited by the human race. I shall not enter into a detail of his proofs, nor undertake their refutation. The exposition we have just made, is sufficient to demonstrate the difference of his opinion with public facts, its contrariety with scripture, and consequently the insufficiency of his proofs. On the whole, he treats this matter as a theological controvertist, rather than as an enlightened philosopher.

Leaving these erroneous principles, he flies to ingenious suppositions, which, although extraordinary, yet have a degree of probability to those who, like him, incline to the enthusiasm of system. He says, that the ancient chaos, the origin of our earth, was the atmosphere of a comet: that the annual motion of the earth began at the time it took its new form, but that its diurnal motion began only when the first man fell. That the ecliptic cut the tropic of cancer, opposite to the terrestrial paradise, which was situated on the north-west side of the frontiers of Assyria: that before the deluge, the year began at the autumnal equinox: that the orbits of the planets, and the earth were then perfect circles. That the deluge began the 18th of November, 2365 of the Julian period, or 2349 years before Christ. That the solar and lunar year were then the same, and that they exactly contained 360 days. That a comet descending in the plane of the ecliptic towards its perihelion, passed near the globe of the earth the same day as the deluge began: that there is a great heat in the internal part of the terrestrial globe, which constantly diffuses itself from the centre to the circumference; that the form of the earth is like that of an egg, the ancient emblem of the globe; that mountains are the lightest part of the earth, &c. He afterwards attributes all the alterations and changes which have happened to the earth, to the universal deluge; then blindly adopts the theory of Woodward, and indiscriminately makes use of all the observations of that author on the pres-

[6] A New Theory of the Earth by William Whiston, 1708.

ent state of the globe; but assumes originality when he speaks of its future state: according to him it will be consumed by fire, and its destruction will be preceded by terrible earthquakes, thunder, and frightful meteors; the Sun and Moon will have an hideous aspect, the heavens will appear to fall, and the flames will be general over all the earth; but when the fire shall have devoured all the impurities it contains; when it shall be vitrified and rendered transparent as crystal, the saints and the blessed spirits will return and take possession of it, and there remain till the day of judgment.

These hypotheses, at the first glance, appear to be rash and extravagant assertions; nevertheless the author has managed them with such address, and treated them with such strength, that they cease to appear absolutely chimerical. He supports his subjects with much science, and it is surprising that, from a mixture of ideas so very absurd, a system could be formed with an air of probability. It has not affected vulgar minds so much as it has dazzled the eyes of the learned, because they are more easily deceived by the glare of erudition, and the power of novel ideas. Mr. Whiston was a celebrated astronomer, in the constant habit of considering the heavens, observing the stars, and contemplating the wonderful course of nature; he could never persuade himself that this small grain of sand, this Earth which we inhabit, occupied more the attention of the Creator than the universe, the vast extent of which contains millions of other Suns and Earths. He pretends, that Moses has not given us the history of the first creation of this globe, but only a detail of the new form that it took when the Almighty turned it from the mass of a comet into a planet, and formed it into a proper habitation for men. Comets are, in fact, subjected to terrible vicissitudes by reason of the eccentricity of their orbits. Sometimes, like that in 1680, it is a thousand times hotter there than red-hot iron; and sometimes a thousand times colder than ice; if they are, therefore, inhabited it must be by strange creatures, of which we can have no conception.

The planets, on the contrary, are places of rest, where the distance of the sun not varying much, the temperature remains nearly the same, and permits different kinds of plants and animals to grow and multiply.

In the beginning God created the world; but, observes our author, the earth was then an uninhabitable comet, suffering alternatively the excess of heat and cold, its liquifying and freezing by turns formed a chaos, or an abyss, surrounded with thick darkness: "and darkness covered the face of the deep," *& tenebræ erant superfaciam abissi.* This chaos was the atmosphere of the comet, a body composed of heterogeneous matters, the centre occupied by spherical, solid, and hot substances, of about two thousand leagues in diameter, round which a very great surface of a thick fluid extended, mixed with an unshapen and confused matter, like the chaos of the ancient *rudis & indigestaque moles.*

This vast atmosphere contained but very few dry, solid, or terrestrial particles, still less aqueous or aerial, but a great quantity of fluid, dense and heavy matters, mixed, agitated and jumbled together in the greatest disorder and confusion. Such was the earth before the six days, but on the first day of the creation, when the eccentrical orbit of the comet had been changed, every thing took its place, and bodies arranged themselves according to the law of gravity, the heavy fluid descended to the lowest places, and left the upper regions to the terrestrial, aqueous and aerial parts; those likewise descended according to their order of gravity; first the earth, then the water, and last of all the air. The immense volume of chaos was thus reduced to a globe of a moderate size, in the centre of which is the solid body that still retains the heat which the sun formerly communicated to it, when it belonged to a comet. This heat may possibly endure six thousand years, since the comet of 1680 required fifty thousand years to cool. Around this solid and burning matter, which occupies the centre of the earth, the dense and heavy fluid which descended the first is to be found, and this is the fluid which forms the great abyss on which the earth is borne, like cork on quicksilver; but as the terrestrial parts were originally mixed with a large quantity of water, in descending they have dragged with them a part of this water, which, not being able to re-ascend after the earth was consolidated, formed a concentrical bed with the heavy fluid which surrounds this hot substance, insomuch that the great abyss is composed of two concentrical orbs, the most internal of which is a heavy fluid, and the other water; the last of which serves for a foundation to the earth. It is from this admirable arrangement, produced by the atmosphere of a comet, that the Theory of the Earth, and the explanation of all its phenomena are to depend.

When the atmosphere of the comet was once disembarrassed from all the solid and terrestrial matters, there remained only the lighter air, through which the rays of the sun freely passed and instantly produced light: "Let there be light, and there was light." The columns which composed the orb of the Earth being formed with such great precipitation is the cause of their different densities: consequently the heaviest sunk deeper into this subterraneous fluid than the lightest; and it is this which has produced the vallies and mountains on the surface of the earth. These inequalities were, before the deluge, dispersed and situated otherwise than they are at present. Instead of the vast valley, which contains the ocean, there were many small divided cavities on the surface of the globe, each of which contained a part of this water; the mountains were also more divided, and did not form chains as at present: nevertheless, the earth contained a thousand times more people, and was a thousand times more fertile; and the life of man and other animals were ten times longer, all which was affected by the internal heat of the earth that proceeded from the centre, and gave birth to a great number of plants and animals, bestowing on them a degree of vigour necessary for them to subsist a long time, and multiply in great abundance. But this heat, by increasing the strength of bodies, unfortunately extended to the heads of men and animals; it augmented their passions; it deprived man of his innocence, and the brute creation of part of their intelligence; all creatures, excepting fish, who inhabited a colder element, felt the effects of this heat, became criminal and merited death. It therefore came, and this universal death happened on Wednesday the 28th of November, by a terrible deluge of forty days and forty nights, and was caused by the tail of another comet which encountered the earth in returning from its perihelion.

The tail of a comet is the lightest part of its atmosphere; it is a transparent mist, a subtile vapour, which the heat of the sun exhales from the body of the comet: this vapour composed of extremely rarefied aqueous and aerial particles, follows the comet when it descends to its perihelion, and precedes when it re-ascends, so that it is always situate opposite to the sun, as if it sought to be in the shade, and avoid the too great heat of that luminary. The column which this vapour forms is often of an immense length, and the more a comet approaches the sun, the longer and more extended is its tail, and as many comets descend below the annual orb of the earth, it is not surprising that the earth is sometimes found surrounded with the vapour of this tail; this is precisely what happened at the time of the deluge. In two hours the tail of a comet will evacuate a quantity of water equal to what is contained in the whole ocean. In short, this tail was what Moses calls the cataracts of Heaven, "and the cataracts of Heaven were opened." The terrestrial globe meeting with the tail of a comet, must, in going its course through this vapour, appropriate to itself a part of the matter which it contains; all which, coming within the sphere of the earth's attraction, must fall on it, and fall in the form of rain, since this tail is partly composed of aqueous vapours. Thus rain may come down in such abundance as to produce an universal deluge the waters of which might easily surmount the tops of the highest mountains. Nevertheless, our author, cautious of not going directly against the letter of holy writ, does not say that this rain was the sole cause of the universal deluge, but takes the water from every place he can find it. The great abyss as we see contains a considerable quantity. The earth, at the approach of the comet, would prove the force of its attraction; and the waters contained in the great abyss would be agitated by so violent a kind of flux and reflux, that the superficial crust would not resist, but split in several places, and the internal waters be dispersed over the surface, "And the fountains of the abyss were opened."

But what became of these waters, which the tail of the comet and great abyss furnished so liberally? our author is not the least embarrassed thereon. As soon as the earth, continuing its course, removed from the comet, the effects of its attraction, the flux and reflux in the great abyss ceased of course, and immediately the upper waters precipitated back with violence by the same roads as they had been forced upon the surface. The great abyss absorbed all the superfluous waters, and was of a sufficient capacity not only to receive its own waters, but also all those which the tail of the comet had left, because during its agitation, and the rupture of its crust, it had enlarged the space by driving out on all sides the earth that surrounded it. It was at this time also the figure of the earth, which till then was spherical, became elliptic. This effect was occasioned by the centrifugal force caused by its diurnal motion, and by the attraction of the comet, for the earth, in passing through the tail of the comet, found itself so placed that it presented the parts of the equator to that planet; and the power of the attraction of the comet, concurring with the centrifugal force of the earth, caused the parts of the equator to be elevated, and that with the more facility as the crust was broken and divided in an infinity of places, and because the flux and reflux of the abyss drove against the equator more violently than elsewhere.

Here then is Mr. Whiston's history of the creation; the causes of the universal deluge; the length of the life of the first men; and the figure of the Earth; all which seem to have cost our author little or no labour; but Noah's ark appears to have greatly disquieted him. In the midst of so terrible a disorder occasioned by the conjunction of the tail of a comet with the waters of the great abyss, in the terrible moments wherein not only the elements of the earth were confused, but when new elements still concurred to augment the chaos, how can it be imagined that the ark floated quietly with its numerous cargo on the top of the waves? Here our author makes great efforts to arrive at and give a physical reason for the preservation of the ark, but which has always appeared to me insufficient, poorly imagined, and but little orthodoxical: I will not here relate it, but only observe how hard it is for a man who has explained objects so great and wonderful, without having recourse to a supernatural power, to be stopt by one particular circumstance; our author, however, chose rather to risk drowning with the ark, than to attribute to the immediate bounty of the Almighty the preservation of this precious vessel.

I shall only make one remark on this system, of which I have made a faithful abridgement: which is, whenever we are rash enough to attempt to explain theological truths by physical reasons, or interpret purely by human views, the divine text of holy writ, or that we endeavour to reason on the will of the Most High, and on the execution of his decrees, we consequently shall involve ourselves in the darkness and chaos of obscurity and confusion, like the author of this system, which, in defiance of its absurdities, has been received with great applause. He neither doubts the truth of the deluge, nor the authenticity of the sacred writ; but as he was less employed with it than with physic and astronomy, he has taken passages of the scripture for physical facts, and the results of astronomical observations; and has so strangely blended the divine knowledge with human science as to give birth to the most extraordinary system that possibly ever was or will be conceived.

ARTICLE III.

FROM THE SYSTEM OF BURNET.[7]

This author is the first who has treated this subject generally and in a systematical matter. He was possessed of much understanding, and was a person well acquainted with the *belles lettres*. His work acquired great reputation, and was criticised by many of the learned, among the rest by Mr. Keil, who has geometrically demonstrated the errors of Mr. Burnet, in a treatise called "Examination of the Theory of the Earth." Mr. Keil also refuted Whiston's system; but he treats the last author very different from the first, and seems even to be of his opinion in several cases, and looks upon the tail of a comet to be a very probable cause for the deluge. But, to return to Burnet, his book is elegantly written; he knew how to paint noble images and magnificent scenes. His plan is great, but the execution is deficient for want of proper materials: his reasoning is good, but his proofs are weak; yet his confidence in his writings is so great, that he frequently causes his readers to pass over his errors.

He begins by telling us, that before the deluge the earth had a very different form from that which it has at present; it was at first, he says, a fluid mass, compounded of matters of all kinds, and all sorts of figures, the heaviest descended towards the centre, and formed a hard and solid body; round which the waters collected, and the air, and all the liquors lighter than water, surmounted them. Between the orb of air and that of water, was an orb of oily matter, but as the air was still very impure, and contained a great quantity of small particles of terrestrial matter, they by degrees descended on the coat of oil, and formed a terrestrial orb blended with earth and oil; and this was the first habitable earth, and the first abode of man. This was an excellent soil, light, and calculated to yield to the tenderness of the first germs. The surface of the terrestrial globe was at first equal, uniform, without mountains, without seas, and without inequalities; but it remained only about sixteen centuries in this state, for the heat of the sun by degrees drying the crust, split it at first on the surface, soon after these cracks penetrated farther and increased so considerably by time, that at length they entirely opened the crust; in an instant the whole earth fell into pieces in the abyss of water it surrounded; and this was the cause of the deluge.

But all these masses of earth, by falling into the abyss, dragged along with them a great quantity of air; these struck against each other, divided, and accumulated so irregularly, that great cavities filled with air were left between them. The

[7] Thomas Burnet. Telluris theoria sacra, orbis nostri originem & mutationes generales, quas aut jam subut, aut olim Subiturus est complectens. Londina, 1681.

waters by degrees opened these cavities, and in proportion as they filled them, the surface of the earth discovered itself in the highest parts; at length water alone remained in the lowest parts; that is to say, the vast vallies which contain the sea. Thus our ocean is a part of the ancient abyss, the rest is entered into the internal cavities with which the ocean communicates. The islands and sea rocks are the small fragments, and continents are the great masses of the old crust. As the rupture and the fall of this crust are made of a sudden, and with confusion, it was not surprising to find eminences, depths, plains, and inequalities of all kinds on the surface of the earth.

ARTICLE IV.

FROM THE SYSTEM OF WOODWARD.

It may be said of this author, that he attempted to raise an immense monument on a less solid base than the moving sand, and to construct a world with dust; for he pretends, that at the time of the deluge a total dissolution of the earth was made. The first idea which presents, after having gone through his book,[8] is, that this dissolution was made by the waters of the great abyss. He asserts, that the abyss where the water was included opened all at once at the command of God, and dispersed over the surface an enormous quantity of water necessary to cover the tops of the highest mountains, and that God suspended the cause of cohesion which reduced all solid bodies into dust, &c. He did not consider that by these suppositions he added other miracles to that of the universal deluge, or at least physical impossibilities, which agree neither with the letter of the holy writ, nor with the mathematical principles of natural philosophy. But as this author has the merit of having collected many important observations, and as he was better acquainted with the materials of which the globe is composed than those who preceded him, his system, although badly conceived, and worse digested, has nevertheless dazzled many people, who, seduced by the truth of some particular circumstances, put confidence in his general conclusions; we shall, therefore, give a short view of his theory, in which, by doing justice to the author's merit, and the exactness of his observations, we shall put the reader in a state of judging of the insufficiency of his system, and of the falsity of some of his remarks. Mr. Woodward speaks of having discovered by his sight that all matters which compose the English earth, from the surface to the deepest places which had been dug, were disposed by beds of strata, and that in a great number of these there were shells and other marine productions; he afterwards adds, that by his correspondents and friends he was assured, that in other countries the earth is composed of the same materials, and that shells are found there, not only in the plains but on the highest mountains, in the deepest quarries, and in an infinity of different places. He perceived their strata to be horizontal and disposed one over the other, as matters are which are transported by the waters, and deposited in form of sediment. These general remarks, which are true, are followed by particular observations, by which he evidently shews, that fossils found incorporated in the strata are real shells and marine productions, not minerals and singular bodies, the sport of nature, &c.

To these observations, though partly made before him, which he has collected and proved, he adds others less exact. He asserts, that all matters of different strata

---

[8] An Essay towards the Natural History of the Earth, &c. by John Woodward.

are placed one on the other in the order of their specific gravity.

This general assertion is not true, for we daily see rocks placed above clay, sand, coal, and bitumen, and which certainly are specifically heavier than either of these latter materials. If, in fact, we found throughout the earth that the first strata was bitumen, then chalk, then marl, clay, sand, stone, marble, and at last metals, so that the composition of the earth exactly followed the law of gravity, there would be an appearance that they might have been precipitated at the same time, which our author asserts with confidence, in spite of the evidence to the contrary; for, without being a naturalist, we need only have our eye-sight to be convinced that heavy strata are often found above lighter, and that consequently these sediments were not precipitated all at one time, but have been brought and deposited successively by the water. As this is the foundation of his system, and is manifestly false, we shall follow it no farther than to show how far an erroneous principle may produce false combinations and erroneous conclusions.

All the matters, says our author, which compose the earth, from the summits of the highest mountains, to the greatest depths of mines, are disposed by strata, according to their specific weights; therefore he concludes the whole has been dissolved and precipitated at one time. But in what manner, and at what time was it dissolved? In water, replies he, and at the time of the deluge. But there is not a sufficient quantity of water on the globe for this to be effected, since there is more land than water, and the bottom of the sea itself is earth. This he admits, but says, there is more water than is requisite at the centre of the earth, that it was only necessary for it to ascend, and possess a power of dissolving every substance but shells, afterwards to find the means for this water to re-enter the abyss, and to make all this agree with the history of the deluge. This then is the system, of which the author does not entertain the least doubt; for when it is opposed to him that water cannot dissolve marble, stone, and metals, especially in forty days, the duration of the deluge, he answers simply, that nevertheless it did happen so. When he is asked, what the virtue of this water of the abyss was, to dissolve all the earth, and at the same time preserve the shells? he says, that he never pretended that this water was a dissolvent; but that it is clear, by facts, that the earth has been dissolved and the shells preserved. When he was evidently shown that if he had no reason to give, or facts to support, for these phenomena, his system was useless, he said, we have only to imagine that, during the deluge, the force of gravity and the coherency of matter ceased on a sudden, and by this supposition the dissolution of the old world would be explained in a very easy and satisfactory manner. But, it was said to him, if the power which holds the parts of matter united was suspended, why were not the shells dissolved as well as all the rest? Here he makes a discourse on the organization of shells and bones of animals, by which he pretends to prove that their texture being fibrous, and different from that of minerals, their power of cohesion was different also; after all, we have, says he, only to suppose that the power of gravity and cohesion did not entirely cease, but that it was only diminished sufficient to disunite all the parts of minerals, and not those of animals. To all this we cannot be prevented from discovering, that our author's philosophy was not equal to his talents for observation; and I do not think it necessary seriously to refute opinions which have no foundation, especially when they have been imagined against the rules of probability, and drawn from consequences contrary to mechanical laws.

ARTICLE V.

EXPOSITION OF SOME OTHER SYSTEMS.

It is plain that the three forementioned hypotheses have much in common with each other. They all agree in this point, that during the deluge the earth changed its form, as well externally as internally; but these speculators have not considered that the earth before the deluge was inhabited by the same species of men and animals, and must necessarily have been nearly such as it is at present. The sacred writings teach us, that before the deluge there were rivers, seas, mountains, and forests. That these rivers and mountains were, for the most part, retained in the same situations; the Tigris and Euphrates were the rivers of the ancient paradise; that the mountain of Armenia, on which the ark rested, was one of the highest mountains in the world at the deluge, as it is at present: that the same plants and animals which exist now, existed then; for we read of the serpent, of the raven, of the crow, and of the dove, which brought the olive branch into the ark. Although Tournefort asserts there are no olive trees for more than 400 miles from Mount Ararat, and passes some absurd jokes thereon[9], it is nevertheless certain there were olives in this neighbourhood at the time of the deluge, since holy writ assures us of it in the most express terms; but it is by no means astonishing that in the space of 4000 years the olive trees should have been destroyed in those quarters, and multiplied in others; it is therefore contrary to scripture and reason, that those authors have supposed the earth was quite different from its present state before the deluge; and this contradiction between their hypothesis and the sacred text, as well as physical truths, must cause their systems to be rejected, if even they should agree with some phenomena. Burnet gives neither observations, nor any real facts, for the support of his system. Woodward has only given us an essay, in which he promised much more than he could perform: his book is a project, the execution of which has not been seen. He has made use of two general observations; the first, that the earth is every where composed of matters which formerly were in a state of fluidity, transported by the waters, and deposited in horizontal strata. The second, that there are abundance of marine productions in most parts of the bowels of the earth. To give a reason for these facts, he has recourse to the universal deluge, or rather it appears that he gives them as proofs of the deluge; but, like Burnet, he falls into evident contradictions, for it is not to be supposed with them that there were no mountains prior to the deluge, since it is expressly stated, that the waters rose fifteen cubits above the tops of the highest mountains. On the other hand, it is not said that these waters destroyed or dissolved these

---

[9] Voyage du Levant, vol. 2, page 336.

mountains; but, on the contrary, these mountains remained in their places, and the ark rested on that which the water first deserted. Besides how can it be imagined that, during the short duration of the deluge, the waters were able to dissolve the mountains and the whole body of the earth? Is it not an absurdity to suppose that in forty days all marble, rocks, stones, and minerals, were dissolved by water? Is it not a manifest contradiction to admit this total dissolution, and at the same time maintain that shells, bones, and marine productions were preserved entire, and resisted that which had dissolved the most solid substances? I shall not therefore hesitate to say, that Woodward, with excellent facts and observations, has formed but a poor and inconsistent system.

Whiston, who came last, greatly enriched the other two, and notwithstanding he gave a vast scope to his imagination has not fallen into contradiction; he speaks of matters not very credible, but they are neither absolutely nor evidently impossible. As we are ignorant of the centre of the earth, he thought he might suppose it was a solid matter, surrounded with a ring of heavy fluid, and afterwards with a ring of water, on which the external crust was sustained; in the latter the different parts of this crust were more or less sunk, in proportion to their relative weights, which produced mountains and inequalities on the surface of the earth. Here, however, this astronomer has committed a mechanical blunder; he did not recollect that the earth, according to this hypothesis, must be an uniform arch, and that consequently it could not be borne on the water it contains, and much less sunk therein. I do not know that there are any other physical errors; but he has made a great number of errors, both in metaphysics and theology. On the whole it cannot be denied absolutely that the earth meeting with the tail of a comet might not be inundated, especially allowing the author that the tail of a comet may contain aqueous vapours; nor can it be denied as an absolute impossibility that the tail of a comet, in returning from its perihelium, might not burn the earth, if we suppose, with Mr. Whiston, that the comet passed very near the sun; it is the same with the rest of the system. But though his ideas are not absolutely impossibilities, there is so little probability to each thing, when taken separately, that the result upon the whole taken together puts it beyond credibility.

The three systems we have spoken of are not the only works which have been composed on the theory of the earth; a Memoir of M. Bourguet appeared in 1729, printed at Amsterdam, with his "Philosophical Letters on the Formation of Salts, &c." in which he gives a specimen of the system he meditated, but which was prevented completion by the death of the author. It is but justice to admit, that no person was more industrious in making observations or collecting facts. To him we owe that great and beautiful observation, the correspondence between the angles of mountains. He presents every thing which he had collected in great order; but with all those advantages, it appears that he has succeeded no better than the rest in making a physical and reasonable history of the changes which had happened to the globe, and that he was very wide from having found the real cause of those effects which he relates. To be convinced of this we need only cast our eyes on the propositions which he deduces from the phenomena, and which ought to serve for the basis of his theory. He says, that the whole globe took its form at one time, and not successively; that its form and disposition prove that it has been in a state of fluidity; that the present state of the earth is very different from that in which it was for many ages after its first formation; that the matter of the globe was at the beginning less dense than since it altered its appearance; that the condensation of its solid parts diminished by degrees with its velocity, so that after having made a number of revolutions on its axis, and round the sun, it found itself on a sudden in a state of dissolution, which destroyed its first structure. This happened about the vernal equinox. That the sea-shells introduced themselves into the dissolved matters; that after this dissolution the earth took the form it now has, and that the fire which directly infused itself therein consumed it by degrees, and it will be one day destroyed by a terrible explosion, accompanied with a general conflagration, which will augment the atmosphere of the globe, and diminish its diameter, and that then the earth, instead of beds of sand or earth, will have only strata of calcined metal and mountains composed of amalgamas of different metals.

This is sufficient to shew the system M. Bourguet meditated; to divine in this manner the past, and predict the future, nearly as others have predicted, does not appear to me to be an effort of judgment: this author had more erudition than sound and general views: he appears to be deficient in that capaciousness of ideas necessary to follow the extent of the subject, and enable him to comprehend the chain of causes and effects.

In the acts of Leipsic, the famous Leibnitz published a scheme of quite a different system, under the title of *Protogaea*. The earth, according to Bourguet and others, must end by fire; according to Leibnitz it began by it, and has suffered many more changes and revolutions than is imagined. The greatest part of the terrestrial matter was surrounded by violent flames at the time when Moses says light was divided from darkness. The planets, as well as the earth, were fixed stars, luminous of themselves. After having burnt a long time, he pretends that they were extinguished for want of combustible matter, and are become opaque bodies. The fire, by melting the matter, produced a vitrified crust, and the basis of all the matter which composes the globe is glass, of which sand and gravel are only fragments. The other kinds of earth are formed from a mixture of this sand, with fixed salts and water, and when the crust cooled, the humid particles, which were raised in form of vapours, refel, and formed the sea. They at first covered the whole surface, and even surmounted the highest mountains. According to this author, the shells, and other wrecks of the sea, which are every where to be found, positively prove that the sea has covered the whole earth; and the great quantity of fixed salts, sand, and other melted and calcined matters, which are included in the bowels of the earth, prove that the conflagration had been general, and that it preceded the existence of the sea. Although these thoughts are void of proofs, they are capital. The ideas have connection, the hypotheses are not impossible, and the consequences that may be drawn therefrom are not contradictory: but the grand defect of this theory is, that it is not applicable to the present state of the earth; it is the past which it explains, and this past is so far back, and has left us so few remains, that we may say what we please of it, and the probability will be in proportion as a man has talents to elucidate what he asserts. To affirm as Whiston has done, that the earth was originally a comet, or, with Leibnitz, that it has been a sun, is saying things equally possible or impossible, and to which it would be ridiculous to apply the rules of probability. To say that the sea formerly covered all the earth, that it surrounded the whole globe, and that it is for this reason shells are every where found, is not paying attention to a very essential point, the unity of the time of the creation; for if that was so, it must necessarily be admitted, that shell-fish, and other inhabitants of the sea, of which we find the remains in the internal part of the earth, existed long before man, and all terrestrial animals. Now, independent of the testimony of holy writ, is it not reasonable to think, that all animals and vegetables are nearly as ancient as each other?

M. Scheutzer, in a Dissertation, addressed to the Academy of Sciences in 1728, attributes, like Woodward, the change, or rather the second formation of the globe, to the universal deluge; to explain that of mountains, he says, that after the deluge, God chusing to return the waters into subterraneous reservoirs, broke and displaced with his all-powerful hand a number of beds, before horizontal, and raised them above the surface of the globe, which was originally level. The whole Dissertation is composed to imply this opinion. As it was requisite these eminences should be of a solid consistence, M. Scheutzer remarks, that God only drew them from places where there were many stones; from hence, says he, it proceeds that those countries, like Switzerland, which are very stony, are also mountainous; and on the contrary, those, as Holland, Flanders, Hungary and Poland, have only sand or clay, even to a very great depth, and are almost entirely without mountains.[10]

This author, more than any other, is desirous of blending Physic with Theology, and though he has given some good observations, the systematical part of his works is still weaker than those who preceded him. On this subject he has even made declamations and ridiculous witticisms, as may be seen in his *Visciam quærelæ*, &c. without speaking of his large work in many folio volumes, *Physica Sacra*, a puerile work, which appears to be composed less for the instruction of men than for the amusement of children.

Steno, and some others, have attributed the cause of the inequalities of the earth to particular inundations, earthquakes, &c. but the effects of these secondary causes have been only able to produce some slight changes. We admit of these causes after the first cause, the motion of the flux and reflux, and of the sea from east to west. Neither Steno, nor the rest, have given theory, nor even any general facts on this matter.[11]

Ray pretends that all mountains have been produced by earthquakes, and he has composed a treatise to prove it; we shall shew under the article of Volcanos what little foundation his opinion is built upon.

We cannot dispense with observing that Burnet, Woodward, Whiston, and most of these other authors, have committed an error which deserves to be cleared up; which is, to have looked upon the deluge as possible by the action of natural causes, whereas scripture presents it to us as produced by the immediate will of

---

[10] See the Hist. of the Acad. 1708, page 32.
[11] See the Diss. de Solido intra Solidum, &c.

God; there is no natural cause which can produce on the whole surface of the earth, the quantity of water required to cover the highest mountains; and if even we could imagine a cause proportionate to this effect, it would still be impossible to find another cause capable of causing the water to disappear: allowing Whiston, that these waters proceeded from the tail of a comet, we deny that any could proceed from the great abyss, or that they all returned into it, since the great abyss, according to him, being surrounded on every side by the crust, or terrestrial orb, it is impossible that the attraction of the comet could cause any motion to the fluids it contained; much less, as he says, a violent flux and reflux; hence there could not be issued from, nor entered into, the great abyss, a single drop of water; and unless it is supposed that the waters which fell from the comet have been destroyed by a miracle, they would still be on the surface of the earth, covering the summits of the highest mountains. Nothing better characterises a miracle, than the impossibility of explaining the effect of it by natural causes. Our authors have made vain efforts to give a reason for the deluge; their physical efforts, and the secondary causes, which they made use of, prove the truth of the fact as reported in the scriptures, and demonstrate that it could only have been performed by the first cause, the will of the Almighty.

Besides, it is certain that it was neither at one time, nor by the effect of the deluge, that the sea left dry these continents we inhabit: for it is certain by the testimony of holy writ, that the terrestrial paradise was in Asia, and that Asia was inhabited before the deluge; consequently the sea, at that time, did not cover this considerable part of the globe. The earth, before the deluge, was nearly as it is at present, and this enormous quantity of water, which divine justice caused to fall on the earth to punish guilty men, in fact, brought death on every creature; but it produced no change on the surface of the earth, it did not even destroy plants which grew upon it, since the dove brought an olive branch to the ark in her beak.

Why, therefore, imagine, as many of our naturalists have done, that this water totally changed the surface of the globe even to a depth of two thousand feet? Why do they desire it to be the deluge which has brought the shells on the earth which we meet with at 7 or 800 feet depth in rocks and marble? Why say, that the hills and mountains were formed at that time? And how can we figure to ourselves, that it is possible for these waters to have brought masses and banks of shells 100 miles long? I see not how they can persist in this opinion, at least, without admitting a double miracle in the deluge; the first, for the augmentation of the waters; and the second, for the transportation of the shells; but as there is only the first which is related in the Bible, I do not see it necessary to make the second an article of our creed.

On the other hand, if the waters of the deluge had retired all at once, they would have carried so great a quantity of mud and other impurities, that the Earth would not have been capable of culture till many ages after this inundation; as is known, by the deluge which happened in Greece, where the overflowed country was totally forsaken, and could not receive any cultivation for more than three centuries.[12] We ought also to look on the universal deluge as a supernatural

---

[12] See Acta erudit, Lepiss, Ann. 1691, page 100.

means of which the Almighty made use for the chastisement of mankind, and not as an effect of a natural cause. The universal deluge is a miracle both in its cause and effects; we see clearly by the scripture that it was designed for the destruction of men and animals, and that it did not in any mode change the earth, since after the retreat of the waters, the mountains, and even the trees, were in their places, and the surface of the earth was proper to receive culture and produce vines and fruits. How could all the race of fish, which did not enter the ark, be preserved, if the earth had been dissolved in the water, or only if the waters had been sufficiently agitated to transport shells from India to Europe, &c.?

Nevertheless, this supposition, that it was the deluge which transported the shells of the sea into every climate, is the opinion, or rather the superstition, of naturalists. Woodward, Scheutzer, and some more, call these petrified shells the remains of the deluge; they look on them as the medals and monuments which God has left us of this terrible event, in order that it never should be effaced from the human race. In short, they have adopted this hypothesis with so much enthusiasm, that they appear only desirous to reconcile holy scripture with their opinion; and instead of making use of their observations, and deriving light therefrom, they envelope themselves in the clouds of a physical theology, the obscurity of which is derogatory to the simplicity and dignity of religion, and only leaves the absurd to perceive a ridiculous mixture of human ideas and divine truths. To pretend to explain the universal Deluge, and its physical causes; to attempt to teach what passed in the time of that great revolution; to divine what were the effects of it; to add facts to those of Holy Writ, to draw consequences from such facts, is only a presumptuous attempt to measure the power of the Most High. The natural wonders which his benevolent hand performs in an uniform and regular manner, are incomprehensible; and by the strongest reason, these wonderful operations and miracles ought to hold us in awful wonder, and in silent adoration.

But they will say, the universal Deluge being a certain fact, is it not permitted to reason on its consequences? It may be so; but it is requisite that you should begin by allowing that the Deluge could not be performed by physical causes; you ought to consider it is an immediate effect of the will of the Almighty; you ought to confine yourselves to know only what the Holy Writ teaches, and particularly not to blend bad philosophy with the purity of divine truth. These precautions, which the respect we owe to the Almighty exacts, being taken, what remains for examination on the subject of the Deluge? Does the Scripture say mountains were formed by the Deluge? No, it says the contrary. Is it said that the agitation of the waters was so great as to raise up shells from the bottom of the sea, and transport them all over the earth? No; the ark floated quietly on the surface of the waters. Is it said, that the earth suffered a total dissolution? None at all: the recital of the sacred historian is simple and true, that of these naturalists complex and fabulous.

ARTICLE VI.

GEOGRAPHY.

The surface of the Earth, like that of Jupiter, is not divided by bands alternative and parallel to the equator; on the contrary, it is divided from one pole to the other, by two bands of earth, and two of sea; the first and principal is the ancient continent, the greatest length of which is found to be in a line, beginning on the east point of the northern part of Tartary, and extending from thence to the land which borders on the gulph of Linchidolkin, where the Muscovites fish for whales; from thence to Tobolski, from Tobolski to the Caspian sea, from the Caspian sea to Mecca, and from Mecca to the western part of the country inhabited by the Galli, in Africa; afterwards to Monoemuci or Monomotapa, and at last to the Cape of Good Hope; this line, which is the greatest length of the old continent, is about 3600 leagues, Paris measure; it is only interrupted by the Caspian and Red seas, the breadths of which are not very considerable, and we must not pay any regard to these interruptions, when it is considered, the surface of the globe is divided only in four parts.

This greatest length is found by measuring the old continent diagonally; for if measured according to the meridians, we shall find that there are only 2500 leagues from the northernmost Cape of Lapland to the Cape of Good Hope; and that the Baltic and Mediterranean cause a much greater interruption than is met with in the other way. With respect to all the other distances that might be measured in the old continent under the same meridian, we shall find them to be much smaller than this; having, for example, only 1800 leagues from the most southern point of the island of Ceylon to the northernmost coast of Nova Zembla. Likewise, if we measure the continent parallel to the equator, we find that the greatest uninterrupted length is found from Trefna, on the western coast of Africa, to Ninpo, on the eastern coast of China, and that it is about 2800 leagues. Another course may be measured from the point of Brittany near Brest, extending to the Chinese Tartary; about 2300 leagues. From Bergen, in Norway, to the coast of Kamschatka, is no more than 1800 leagues. All these lines have much less length than the first, therefore the greatest extent of the old continent, is, in fact, from the eastern point of Tartary to the Cape of Good Hope, that is about 3600 leagues.

There is so great an equality of surface on each side of this line, which is also the longest, that there is every probability to suppose it really divides the contents of the ancient continent; for in measuring on one side is found 2,471,092-3/4 square leagues, and on the other 2,469,687.

Agreeable to this, the old continent consists of about 4,940,780 square leagues, which is nearly one-fifth of the whole surface of the globe, and has an inclination towards the equator of about 30 degrees.

The greatest length of the new continent may be taken in a line from the mouth of the river Plata to the lake of Assiniboils. From the former it passes to the lake Caracara; from thence to Mataguais, Pocona, Zongo, Mariana, Morua, St. Fe, and Carthagena; it then proceeds through the gulph of Mexico, Jamaica, and Cuba, passes along the peninsula of Florida, through Apolache, Chicachas, and from thence to St. Louis, Fort le Suer, and ends on the borders of lake Assiniboils; the whole extent of which is still unknown.

This line, which is interrupted only by the Mexican gulph (which must be looked upon as a mediterranean sea) may be about 2500 leagues long, and divides the new continent into nearly two equal parts, the left of which contains about 1,069,286-5/6 leagues square, and that on the right about 1,070,926-1/12; this line, which forms the middle of the band of the new continent, is inclined to the equator about 30 degrees, but in an opposite direction, for that of the old continent extends from the north-east to the south-west, and that of the new continent from the north-west to the south-east. All those lands together of the old and new continent, make about 7,080,993 leagues square, which is not near the third of the whole surface, which contains 25 millions of square leagues.

It must be remarked, that these two lines, which divide the continents into two equal parts, both terminate at the same degree of southern and northern latitude, and that the two continents make opposite projections, which exactly face each other; to wit, the coasts of Africa, from the Canary islands to the coasts of Guinea, and those of America from Guiana to the mouth of Rio Janeiro.

It appears, therefore, that the most ancient land of the globe, is on the two sides of these lines, at the distance of from 2 to 250 leagues on each side. By following this idea, which is founded on the observations before related, we shall find in the old continent that the most ancient lands of Africa are those which extend from the Cape of Good Hope to the Red Sea, as far as Egypt, about 500 leagues broad, and that, consequently, all the western coasts of Africa, from Guinea to the straits of Gibraltar, are the newest lands. So likewise we shall discover that in Asia, if we follow the line on the same breadth, the most ancient lands are Arabia Felix and Deserta, Persia, Georgia, Turcomania, part of Tartary, Circassia, part of Muscovy, &c. that consequently Europe, and perhaps also China, and the eastern part of Tartary, are more modern. In the new continent we shall find the Terra Magellanica, the eastern part of Brasil, the country of the Amazons, Guiana, and Canada, to be the new lands, in comparison with Peru, Terra Firma, the islands in the gulph of Mexico, Florida, the Mississippi, and Mexico.

To these observations we may add two very remarkable facts, the old and new continent are almost opposite each other; the old is more extensive to the north of the equator than the south; the new is more to the south than the north. The centre of the old continent is in the 16th or 18th degree of north latitude, and the centre of the new is in the 16th or 18th degree south latitude, so that they seem to be made to counterbalance each other. There is also a singular connexion between the two continents, although it appears to be more accidental than those which I have spoken of, which is, that if the two continents were divided into two parts, all four would be surrounded by the sea, if it were not for the two small isthmuses, Suez and Panama.

This is the most general idea which an attentive inspection of the globe furnishes us with, on the division of the earth. We shall abstain from forming hypotheses thereon, and hazarding reasonings which might lead into false conclusions; but no one as yet having considered the division of the globe under this point of view, I shall submit a few remarks. It is very singular that the line which forms the greatest length of the terrestrial continents divides them also into two equal parts; it is no less so that these two lines commence and end at the same degrees of latitude, and are both alike inclined to the equator. These relations may belong to some general conclusions, but of which we are ignorant. The inequalities in the figure of the two continents we shall hereafter examine more fully: it is sufficient here to observe, that the most ancient countries are the nearest to these lines, and are the highest; that the more modern lands are the farthest, and also the lowest. Thus in America, the country of the Amazons, Guiana and Canada will be the most modern parts; by casting our eyes on the map of this country we see the waters on every side, and that they are divided by numberless lakes and rivers, which also indicates that these lands are of a late formation; while on the other hand Peru and Mexico are high mountains, and situate at no great distance from the line that divides the continent, which are circumstances that seem to prove their antiquity. Africa is very mountainous, and that part of the world is also very ancient. There are only Egypt, Barbary, and the western coasts of Africa, as far as Senegal, in this part of the globe, which can be looked upon as modern countries. Asia is an old land, and perhaps the most ancient of all, particularly Arabia, Persia, and Tartary; but the inequalities of this vast part of the globe, as well as those of Europe, we will consider in a separate article. It might be said in general, that Europe is a new country, and such position would be supported both by the universal traditions relative to the emigrations of different people, and the origin of arts and sciences. It is not long since it was filled with morasses, and covered with forests, whereas in the land anciently inhabited, there are but few woods, little water, no morasses, much land, and a number of mountains, whose summits are dry and barren; for men destroy the woods, drain the waters, confine rivers, dry up morasses, and in time give a different appearance to the face of the earth, from that, of uninhabited or newly-peopled countries.

The ancients were acquainted with but a small part of the globe. All America, the Magellanic, and a great part of the interior of Africa, was entirely unknown to them. They knew not that the torrid zone was inhabited, although they had navigated around Africa, for it is 2200 years since Neco, king of Egypt, gave vessels to the Phenicians, who sailed along the Red Sea, coasted round Africa, doubled the Cape of Good Hope, and having employed two years in this voyage, the third year they entered the straits of Gibraltar.[13] The ancients were unacquainted with the

---

[13] Vide Herodotus, lib. iv.

property of the loadstone, if turned towards the poles, although they knew that it attracted iron. They were ignorant of the general cause of the flux and reflux of the sea, nor were they certain the ocean surrounded the globe; some indeed suspected it might be so, but with so little foundation, that no one dared to say, or even conjecture, it was possible to make a voyage round the world. Magellan was the first who attempted it in the year 1519, and accomplished the great voyage in 1124 days. Sir Francis Drake was the second in 1577, and he performed it in 1056 days; afterwards Thomas Cavendish made this great voyage in 777 days, in the year 1586. These celebrated navigators were the first who demonstrated physically the sphericity and the extent of the earth's circumference; for the ancients had no conception of the extent of this circumference, although they had travelled a great deal. The trade winds, so useful in long voyages, were also unknown to them; therefore we must not be surprised at the little progress they made in geography. Notwithstanding the knowledge we have acquired by the aid of mathematical sciences, and the discovery of navigators, many things remain still unsettled, and vast countries undiscovered. Almost all the land on the side of the Atlantic pole is unknown to us; we only know that there is some, and that it is separated from all the other continents by the ocean. Much land also remains to be discovered on the side of the Arctic pole, and it is to be regretted that for more than a century the ardour of discovering new countries is extremely abated. European governments seem to prefer, and possibly with reason, increasing the value of those countries we are acquainted with to the glory of conquering new ones.

Nevertheless, the discovery of the southern continent would be a great object of curiosity, and might be useful. We have discovered only some few of its coasts; those navigators who have attempted this discovery, have always been stopt by the ice. The thick fogs, which are in those latitudes, is another obstacle; yet, in defiance of these inconveniencies, it is probable that by sailing from the Cape of Good Hope at different seasons, we might at last discover a part of these lands, which hitherto make a separate world.

There is another method, which possibly might succeed better. The ice and fogs having hitherto prevented the discovery, might it not be attempted by the Pacific Sea; sailing from Baldivia, or any other port on the coast of Chili, and traversing this sea under the 50th degree south latitude? There is not the least appearance that this navigation is perilous, and it is probable would be attended with the discovery of new countries; for what remains for us to know on the coast of the southern pole, is so considerable, that we may estimate it at a fourth part of the globe, and of course may contain a continent, as large as Europe, Asia, and Africa, all together.

As we are not at all acquainted with this part of the globe, we cannot justly know the proportion between the surface of the earth and that of the sea; only as much as may be judged by inspection of what is known, there is more sea than land.

If we would have an idea of the enormous quantity of water which the sea contains, we must suppose a medium depth, and by computing it only at 200 fathom, or the sixth part of a league, we shall find that there is sufficient to cover the whole globe to the height of 600 feet of water, and if we would reduce this water into one mass, it would form a globe of more than 60 miles diameter.

Navigators pretend, that the latitudes near the south pole are much colder than those of the north, but there is no appearance that this opinion is founded on truth, and probably has been adopted, because ice is found in latitudes where it is scarcely ever seen in the southern seas; but that may proceed from some particular cause. We find no ice in April on this side 67 and 68 degrees northern latitude: and the savages of Arcadia and Canada say, when it is not all melted in that month, it is a sign the rest of the year will be cold and rainy. In 1725 there may be said to have been no summer, it rained almost continually; and the ice of the northern sea was not only not melted in April in the 67th degree, but even it was found the 15th of June towards the 41st and 42d degree[14].

A great quantity of floating ice appears in the northern sea, especially at some distance from land. It comes from the Tartarian sea into that of Nova Zembla, and other parts of the Frozen Ocean. I have been assured by people of credit, that an English Captain, named Monson, instead of seeking a passage between the northern land to go to China, directed his course strait to the pole, and had approached it within two degrees; that in this course he had found an open sea, without any ice, which proves that the ice is formed near land, and never in open sea; for if we should suppose, against all probability, that it might be cold enough at the pole to freeze over the surface of the sea, it is still not conceivable how these enormous floating mountains of ice could be formed, if they did not find a fixed point against land, from whence afterwards they were loosened by the heat of the sun. The two vessels which the East India Company sent, in 1739, to discover land in the South Seas, found ice in the latitude of 47 and 48 degrees, but this ice was not far from shore, that being in sight although they were unable to land. This must have been separated from the adjoining lands of the south pole, and it may be conjectured that they follow the course of some great rivers, which water the unknown land, the same as the Oby, Jenisca, and other great floods, which fall into the North Seas, carry with them the ice, which, during the greatest part of the year, stops up the straits of Waigat, and renders the Tartarian sea unnavigable by this course; whereas beyond Nova Zembla, and nearer the poles, where there are few rivers, and but little land, ice is not so frequently met with, and the sea is more navigable; so that if they would still attempt the voyage to China and Japan by the North Seas, we should possibly, to keep clear from the land and ice, shape our course to the pole, and seek the open seas, where certainly there is but little or no ice; for it is known that salt water can, without freezing, become colder than fresh water when frozen, and consequently the excessive cold of the pole may possibly render the sea colder than the ice, without the surface being frozen: so much the more as at 80 or 82 degrees, the surface of the sea, although mixed with much snow and fresh water, is only frozen near the shore. By collecting the testimonies of travellers, on the passage from Europe to China, it appears that one does exist by the north sea; and the reason it has been so often attempted in vain is, because they have always feared to go sufficiently far from land, and approach the pole.

---

[14] See the Hist. of the Acad. Ann. 1725.

Captain William Barents, who, as well as others, run aground in his voyage, yet did not doubt but there was a passage, and that if he had gone farther from shore, he should have found an open sea free from ice. The Russian navigators, sent by the Czar to survey the north seas, relate that Nova Zembla is not an island, but belonging to the continent of Tartary, and that to the north of it is a free and open sea. A Dutch navigator asserts, that the sea throws up whales on the coasts of Corea and Japan, which have English and Dutch harpoons on their backs. Another Dutchman has pretended to have been at the pole, and asserts it is as warm there as it is at Amsterdam in the middle of the summer. An Englishman, named Golding, who made more than thirty voyages to Greenland, related to King Charles II. that two Dutch vessels with which he had sailed, having found no whales on the coast of the island of Edges, resolved to proceed farther north, and that upon their return at the expiration of fifteen days, they told him that they had been as far as 89 degrees latitude (within one degree of the pole), and that they found no ice there, but an open deep sea like that of the Bay of Biscay, and that they shewed him the journals of the two vessels, as a proof of what they affirmed. In short, it is related in the Philosophical Transactions that two navigators, who had undertaken the discovery of this passage, shaped a course 300 leagues to the east of Nova Zembla, but that the East India Company, who thought it their interest this passage should not be discovered, hindered them from returning[15]. But the Dutch East India Company thought, on the contrary, that it was their interest to find this passage; having attempted it in vain on the side of Europe, they sought it by that of Japan, and they would probably have succeeded, if the Emperor of Japan had not forbidden all strangers from navigating on the side of the land of Jesso. This passage, therefore, cannot be found but by sailing to the pole, beyond Spitzbergen, or by keeping the open sea between Nova Zembla and Spitzbergen under the 79th degree of latitude. We need not fear to find it frozen even under the pole itself, for reasons we have alledged; in fact, there is no example of the sea being frozen at a considerable distance from the shore; the only example of a sea being frozen entirely over, is that of the Black Sea, which is narrow, contains but little salt, and receives a number of rivers from the northern countries, and which bring ice with them: and if we may credit historians, it was frozen in the time of the Emperor Copronymus, thirty cubits deep, without reckoning twenty cubits of snow above the ice. This appears to be exaggerated, but it is certain that it freezes almost every winter; whereas the open seas, a thousand leagues nearer the pole, do not freeze at all: this can only proceed from the saltness, and the little ice which they receive, in comparison with that transported into the Black Sea.

This ice, which is looked upon as a barrier that opposes the navigation near the poles, and the discovery of the southern continent, proves only that there are large rivers adjacent to the places where it is met with; and indicates also there are vast continents from whence these rivers flow; nor ought we to be discouraged at the sight of these obstacles; for if we consider, we shall easily perceive, this ice must be confined to some particular places; that it is almost impossible that it should occupy the whole circle which encompasses, as we suppose, the southern continent, and therefore we should probably succeed if we were to direct our

[15] See the collection of Northern Voyages, page 200.

course towards some other point of this circle. The description which Dampier and some others have given of New Holland, leads us to suspect that this part of the globe is perhaps a part of the southern lands, and is a country less ancient than the rest of this unknown continent. New Holland is a low country, without water or mountains, but thinly inhabited, and the natives without industry; all this concurs to make us think that they are in this continent nearly what the savages of Amaconia or Paraguais are in America. We have found polished men, empires, and kings, at Peru and Mexico, which are the highest, and consequently the most ancient countries of America. Savages, on the contrary, are found in the lowest and most modern countries; therefore we may presume that we should also find men united by the bands of society in the upper countries, from whence these great rivers, which bring this prodigious ice to the sea, derive their sources.

The interior parts of Africa are unknown to us, almost as much as they were to the ancients: they had, like us, made the tour of that vast peninsula, but they have left us neither charts, nor descriptions of the coasts. Pliny informs us, that the tour of Africa was made in the time of Alexander the Great, that the wrecks of some Spanish vessels had been discovered in the Arabian sea, and that Hanno, a Carthaginian general, had made a voyage from Gades to the Arabian sea, and that he had written a relation of it. Besides that, he says Cornelius Nepos tells us that in his time one Eudoxus, persecuted by the king Lathurus, was obliged to fly from his country; that departing from the Arabian gulph, he arrived at Gades, and that before this time they traded from Spain to Ethiopia by sea[16]. Notwithstanding these testimonies of the ancients, we are persuaded that they never doubled the Cape of Good Hope, and the course which the Portuguese took the first to go to the East-Indies, was looked upon as a new discovery; it will not perhaps, therefore, be deemed amiss to give the belief of the 9th century on this subject.

"In our time an entire new discovery has been made, which was wholly unknown to those who lived before us. No one thought, or even suspected, that the sea, which extends from India to China, had a communication with the Syrian sea. We have found, according to what I have learnt, in the sea Roum, or Mediterranean, the wreck of an Arabian vessel, shattered to pieces by the tempest, some of which were carried by the wind and waves to the Cozar sea, and from thence to the Mediterranean, and was at length thrown on the coast of Syria. This proves that the sea surrounds China and Cila, the extremity of Turqueston and the country of the Cozars; that it afterwards flows by the strait till it has washed the coast of Syria. The proof is drawn from the construction of the vessel; for no other vessels but those of Siraf are built without nails, which, as was the wreck we speak of, are joined together in a particular manner, as if they were sewed. Those, of all the vessels of the Mediterranean and of the coast of Syria, are nailed and not joined in this manner[17]."

To this the translator of this ancient relation adds.—

"Abuziel remarks, as a new and very extraordinary thing, that a vessel was

---

[16] Vide Pliny, Hist. Nat. Vol. I. lib. 2.
[17] See the ancient relations of travels by land to China, page 53 and 54.

carried from the Indian sea, and cast on the coasts of Syria. To find a passage into the Mediterranean, he supposes there is a great extent above China, which has a communication with the Cozar sea, that is, with Muscovia. The sea which is below Cape Current, was entirely unknown to the Arabs, by reason of the extreme danger of the navigation, and from the continent being inhabited by such a barbarous people, that it was not easy to subject them, nor even to civilize them by commerce. From the Cape of Good Hope to Soffala, the Portuguese found no established settlement of Moors, like those in all the maritime towns as far as China, which was the farthest place known to geographers; but they could not tell whether the Chinese sea, by the extremity of Africa, had a communication with the sea of Barbary, and they contented themselves with describing it as far as the coast of Zing, or Caffraria. This is the reason why we cannot doubt but that the first discovery of the passage of this sea, by the Cape of Good Hope, was made by the Europeans, under the conduct of Vasco de Gama, or at least some years before he doubled the Cape, if it is true there are marine charts of an older date, where the Cape is called by the name of Frontiera du Africa. Antonio Galvin testifies, from the relation of Francisco de Sousa Tavares, that, in 1528, the Infant Don Ferdinand shewed him such a chart, which he found in the monastery of Acoboca, dated 120 years before, copied perhaps from that said to be in the treasury of St. Mark, at Venice, which also marks the point of Africa, according to the testimony of Ramusio, &c."

The ignorance of those ages, on the subject of the navigation around Africa, will appear perhaps less singular than the silence of the editor of this ancient relation on the subject of the passages of Herodotus, Pliny, &c. which we have quoted, and which proves the ancients had made the tour of Africa.

Be it as it may, the African coasts are now well known; but whatever attempts have been made to penetrate into the inner parts of the country, we have not been able to attain sufficient knowledge of it to give exact relations[18]. It might, nevertheless, be of great advantage, if we were, by Senegal, or some other river, to get farther up the country and establish settlements, as we should find, according to all appearances, a country as rich in precious mines as Peru or the Brazils. It is perfectly known that the African rivers abound with gold, and as this country is very mountainous, and situated under the equator, it is not to be doubted but it contains, as well as America, mines of heavy metals, and of the most compact and hard stones.

The vast extent of north and east Tartary has only been discovered in these latter times. If the Muscovite maps are just, we are at present acquainted with the coasts of all this part of Asia; and it appears that from the point of eastern Tartary to North America, it is not more than four or five hundred leagues: it has even been pretended that this tract was much shorter, for in the Amsterdam Gazette, of the 24th of January, 1747, it is said, under the article of Petersburgh, that Mr. Stalleravoit had discovered one of these American islands beyond Kamschatca, and demonstrated that we might go thither from Russia by a shorter tract. The Jesuits,

---

[18] Since this time, however, great discoveries, have been made; Mons. Vaillant has given a particular description of the country from the Cape to the borders of Caffraria; and much information has also been acquired by the Society for Asiatic Researches.

and other missionaries, have also pretended to have discovered savages in Tartary, whom they had catechised in America, which should in fact suppose that passage to be still shorter[19]. This author even pretends, that the two continents of the old and new world join by the north, and says, that the last navigations of the Japanese afford room to judge, that the tract of which we have spoken is only a bay, above which we may pass by land from Asia to America. But this requires confirmation, for hitherto it has been thought that the continent of the north pole is separated from the other continents, as well as that of the south pole.

Astronomy and Navigation are carried to so high a pitch of perfection, that it may reasonably be expected we shall soon have an exact knowledge of the whole surface of the globe. The ancients knew only a small part of it, because they had not the mariner's compass. Some people have pretended that the Arabs invented the compass, and used it a long time before we did, to trade on the Indian sea, as far as China; but this opinion has always appeared destitute of all probability; for there is no word in the Arab, Turkish, or Persian languages, which signifies the compass; they make use of the Italian word Bossola; they do not even at present know how to make a compass, nor give the magnetical quality to the needle, but purchase them from the Europeans. Father Maritini says, that the Chinese have been acquainted with the compass for upwards of 3000 years; but if that was the case, how comes it that they have made so little use of it? Why did they, in their voyages to Cochinchina, take a course much longer than was necessary? And why did they always confine themselves to the same voyages, the greatest of which were to Java and Sumatra? And why did not they discover, before the Europeans, an infinity of fertile islands, bordering on their own country, if they had possessed the art of navigating in the open seas? For a few years after the discovery of this wonderful property of the loadstone, the Portuguese doubled the Cape of Good Hope, traversed the African and Indian seas, and Christopher Columbus made his voyage to America.

By a little consideration, it was easy to divine there were immense spaces towards the west; for, by comparing the known part of the globe, as for example, the distance of Spain to China, and attending to the revolution of the Earth and Heavens, it was easy to see that there remained a much greater extent towards the west to be discovered, than what they were acquainted with towards the east. It, therefore, was not from the defect of astronomical knowledge that the ancients did not find the new world, but only for want of the compass. The passages of Plato and Aristotle, where they speak of countries far distant from the Pillars of Hercules, seem to indicate that some navigators had been driven by tempest as far as America, from whence they returned with much difficulty; and it may be conjectured, that if even the ancients had been persuaded of the existence of this continent, they would not have even thought it possible to strike out the road, having no guide nor any knowledge of the compass.

I own, that it is not impossible to traverse the high seas without a compass, and that very resolute people might have undertaken to seek after the new world by conducting themselves simply by the stars. The Astrolabe being known to the

---

[19] See the Hist. of New France, by the Pere Charlevoix. Vol. III. page 30 and 31.

ancients, it might strike them they could leave France or Spain, and sail to the west, by keeping the polar star always to the right, and by frequent soundings might have kept nearly in the same latitude; without doubt the Carthaginians, of whom Aristotle makes mention, found the means of returning from these remote countries by keeping the polar star to the left; but it must be allowed that a like voyage would be looked upon as a rash enterprize, and that consequently we must not be astonished that the ancients had not even conceived the project.

Previous to Christopher Columbus's expedition, the Azores, the Canaries, and Madeira were discovered. It was remarked, that when the west winds lasted a long time, the sea brought pieces of foreign wood on the coast of these islands, canes of unknown species, and even dead bodies, which by many marks were discovered to be neither European nor African. Columbus himself remarked, that on the side of the west certain winds blew only a few days, and which he was persuaded were land winds; but although he had all these advantages over the ancients, and the knowledge of the compass, the difficulties still to conquer were so great, that there was only the success he met with which could justify the enterprise. Suppose, for a moment, that the continent of the new world had been 1000 or 1500 miles farther than it in fact is, a thing with Columbus could neither know nor foresee, he would not have arrived there, and perhaps this great country might still have remained unknown. This conjecture is so much the better founded, as Columbus, although the most able navigator of his time, was seized with fear and astonishment in his second voyage to the new world; for as in his first, he only found some islands, he directed his course more to the south to discover a continent, and was stopt by currents, the considerable extent and direction of which always opposed his course, and obliged him to direct his search to the west; he imagined that what had hindered him from advancing on the southern side was not currents, but that the sea flowed by raising itself towards the heavens, and that perhaps both one and the other touched on the southern side. True it is, that in great enterprises the least unfortunate circumstance may turn a man's brain, and abate his courage.

ARTICLE VII.

ON THE PRODUCTION OF THE STRATA, OR BEDS OF EARTH.

We have shewn, in the first article, that by virtue of the mutual attraction between the parts of matter, and of the centrifugal force, which results from its diurnal rotation, the earth has necessarily taken the form of a spheroid, the diameters of which differ about a 230th part, and that it could only proceed from the changes on the surface, caused by the motion of the air and water, that this difference could become greater, as is pretended to be the case from the measures taken under the equator, and within the polar circle. This figure of the earth, which so well agrees with hydrostatical laws, and with our theory, supposes the globe to have been in a state of liquefaction when it assumed its form, and we have proved that the motions of projection and rotation were imprinted at the same time by a like impulsion. We shall the more easily believe that the earth has been in a state of liquefaction produced by fire, when we consider the nature of the matters which the globe incloses, the greatest part of which are vitrified or vitrifiable; especially when we reflect on the impossibility there is that the earth should ever have been in a state of fluidity, produced by the waters; since there is infinitely more earth than water, and that water has not the power of dissolving stone, sand, and other matters of which the earth is composed.

It is plain then that the earth took its figure at the time when it was liquefied by fire: by pursuing our hypothesis it appears, that when the sun quitted it, the earth had no other form than that of a torrent of melted and inflamed vapour matter; that this torrent collected itself by the mutual attraction of its parts, and became a globe, to which the rotative motion gave the figure of a spheroid; and when the earth was cooled, the vapours, which were first extended like the tails of comets, by degrees condensed and fell upon the surface, depositing, at the same time, a slimy substance mixed with sulphurous and saline matters, a part of which, by the motion of the waters, was swept into the perpendicular cracks, where it produced metals, while the rest remained on the surface, and produced that reddish earth which forms the first strata; and which, according to different places, is more or less blended with animal and vegetable particles, so reduced that the organization is no longer perceptible.

Therefore, in the first state of the earth, the globe was internally composed of vitrified matter, as I believe it is at present, above which were placed those bodies the fire had most divided, as sand, which are only fragments of glass; and above these, pumice stones and the scoria of the vitrified matter, which formed the various clays; the whole was covered with water 5 or 600 feet deep, produced by the condensation of the vapours, when the globe began to cool. This water every where deposited a muddy bed, mixed with waters which sublime and exhale by the fire; and the air was formed of the most subtile vapours, which, by their lightness, disengaged themselves from the waters, and surmounted them.

Such was the state of the globe when the action of the tides, the winds, and the heat of the sun, began to change the surface of the earth. The diurnal motion, and the flux and reflux, at first raised the waters under the southern climate, which carried with them mud, clay, and sand, and by raising the parts of the equator, they by degrees perhaps lowered those of the poles about two leagues, as we before mentioned; for the waters soon reduced into powder the pumice stones and other spongeous parts of the vitrified matter that were at the surface, they hollowed some places, and raised others, which in course of time became continents, and produced all the inequalities, and which are more considerable towards the equator than the poles; for the highest mountains are between the tropics and the middle of the temperate zones, and the lowest are from the polar circle to the poles; between the tropics are the Cordeliers, and almost all the mountains of Mexico and Brazil, the great and little Atlas, the Moon, &c. Beside the land which is between the tropics, from the superior number of islands found in those parts, is the most unequal of all the globe, as evidently is the sea.

However independent my theory may be of that hypothesis of what passed at the time of the first state of the globe, I refer to it in this article, in order to shew the connection and possibility of the system which I endeavoured to maintain in the first article. It must only be remarked, that my theory does not stray far from it, as I take the earth in a state nearly similar to what it appears at present, and as I do not make use of any of the suppositions which are used on reasoning on the past state of the terrestrial globe. But as I here present a new idea on the subject of the sediment deposited by the water, which, in my opinion, has perforated the upper bed of earth, it appears to me also necessary to give the reason on which I found this opinion.

The vapours which rise in the air produce rain, dew, aerial fires, thunder, and other meteors. These vapours are therefore blended with aqueous, aerial, sulphurous and terrestrial particles, &c. and it is the solid and earthy particles which form the mud or slime we are now speaking of. When rain water is suffered to rest, a sediment is formed at bottom; and having collected a quantity, if it is suffered to stand and corrupt, it produces a kind of mud which falls to the bottom of the vessel. Dew produces much more of this mud than rain water, which is greasy, unctuous, and of a reddish colour.

The first strata of the earth is composed of this mud, mixed with perished vegetable or animal parts, or rather stony and sandy particles. We may remark that almost all land proper for cultivation is reddish, and more or less mixed with these different matters; the particles of sand or stone found there are of two kinds, the one coarse and heavy, the other fine and sometimes impalpable. The largest comes from the lower strata loosened in cultivating the earth, or rather the upper mould, by penetrating into the lower, which is of sand and other divided matters, and forms those earths we call fat and fertile. The finer sort proceeds from the air, and falls with dew and rain, and mixes intimately with the soil. This is properly the residue of the powder, which the wind continually raises from the surface of the earth, and which falls again after having imbibed the humidity of the air. When the earth predominates, and the stony and sandy parts are but few, the earth is then reddish and fertile: if it is mixed with a considerable quantity of perished animal or vegetable substances, it is blackish, and often more fertile than the first; but if the mould is only in a small quantity, as well as the animal or vegetable parts, the earth is white and sterile, and when the sandy, stony, or cretaceous parts which compose these sterile lands, are mixed with a sufficient quantity of perished animal or vegetable substances, they form the black and lighter earths, but have little fertility; so that according to the different combinations of these three different matters, the land is more or less fecund and differently coloured.

To fix some ideas relative to these stratas; let us take, for example, the earth of Marly-la-ville, where the pits are very deep: it is a high country, but flat and fertile, and its strata lie arranged horizontally. I had samples brought me of all these strata which M. Dalibard, an able botanist, versed in different sciences, had dug under his inspection; and after having proved the matters of which they consisted in aquafortis, I formed the following table of them.

The state of the different beds of earth, found at Marly-la-ville, to the depth of 100 feet.

	Feet.	In.
1. A free reddish earth, mixed with much mud, a very small quantity of vitrifiable sand, and somewhat more of calcinable sand	13	0
2. A free earth mixed with gravel, and a little more vitrifiable sand	2	6
3. Mud mixed with vitrifiable sand in a great quantity, and which made but very little effervescence with aquafortis	3	0
4. Hard marl, which made a very great effervescence with aquafortis	2	0
5. Pretty hard marl stone	4	0
6. Marl in powder, mixed with vitrifiable sand	5	0
7. Very fine vitrified sand	1	6
8. Marl very like earth mixed with a very little vitrifiable sand	3	6
9. Hard marl, in which was real flint	3	6
10. Gravel, or powdered marl	1	0
11. Eglantine, a stone of the grain and hardness of marble, and sonorous	1	6
12. Marly gravel	1	6
13. Marl in hard stone, whose grain was very fine	1	6

14. Marl in stone, whose grain was not so fine	1	6
15. More grained and thicker marl	2	6
16. Very fine vitrifiable sand, mixed with fossil sea-shells, which had no adherence with the sand, and whose colours were perfect	1	6
17. Very small gravel, or fine marl powder	2	0
18. Marl in hard stone	3	6
19. Very coarse powdered marl	1	6
20. Hard and calcinable stone, like marble	1	0
21. Grey vitrifiable sand mixed with fossil shells, particularly oysters and muscles which have no adherence with the sand, and which were not petrified	3	0
22. White vitrifiable sand mixed with similar shells	2	0
23. Sand streaked red and white, vitrifiable and mixed with the like shells	1	0
24. Larger sand, but still vitrifiable and mixed with the like shells	1	0
25. Fine and vitrifiable grey sand mixed with the like shells	8	6
26. Very fine fat sand, with only a few shells	3	0
27. Brown free stone	3	0
28. Vitrifiable sand, streaked red and white	4	0
29. White vitrifiable sand	3	6
30. Reddish vitrifiable sand	15	0
	—————	—————
Total depth	101	0
	—————	—————

I have before said that I tried all these matters in aquafortis, because where the inspection and comparison of matters with others that we are acquainted with is not sufficient to permit us to denominate and range them in the class which they belong, there is no means more ready, nor perhaps more sure, than to try by aquafortis the terrestrial or lapidific matter: those which acid spirits dissolve immediately with heat and ebullition, are generally calcinable, and those on which they make no impression are vitrifiable.

By this enumeration we perceive, that the soil of Marly-la-ville was formerly the bottom of the sea, which has been raised above 75 feet, since we find shells at that depth below the surface. Those shells have been transported by the motion of the water, at the same time as the sand in which they are met with, and the whole of the upper strata, even to the first, have been transported after the same manner by the motion of the water, and deposited in form of a sediment; which we cannot doubt, as well by reason of their horizontal position, as of the different beds of sand mixed with shells and marl, the last of which are only the fragments of the shells. The last stratum itself has been formed almost entirely by the mould we have spoken of, mixed with a small part of the marl which was at the surface.

I have chosen this example, as the most disadvantageous to my theory, because it at first appears very difficult to conceive that the dust of the air, rain and dew, could produce strata of free earth thirteen feet thick; but it ought to be observed, that it is very rare to find, especially in high lands, so considerable a thickness of cultivateable earth; it is generally about three or four feet, and often not more than one. In plains surrounded with hills, this thickness of good earth is the greatest, because the rain loosens the earth of the hills, and carries it into the vallies; but without supposing any thing of that kind, I find that the last strata formed by the waters are thick beds of marl. It is natural to imagine that the upper stratum had, at the beginning, a still greater thickness, besides the thirteen feet of marl, when the sea quitted the land and left it naked. This marl, exposed to the air, melted with the rain; the action of the air and heat of the sun produced flaws, and reduced it into powder on the surface; the sea would not quit this land precipitately, but sometimes cover it, either by the alternative motion of the tides, or by the extraordinary elevation of the waters in foul weather, when it mixed with this bed of marl, mud, clay, and other matters. When the land was raised above the waters, plants would begin to grow, and it was then that the dust in the rain or dew by degrees added to its substance and gave it a reddish colour; this thickness and fertility was soon augmented by culture; by digging and dividing its surface, and thus giving to the dust, in the dew or rain, the facility of more deeply penetrating it, which at last produced that bed of free earth thirteen feet thick.

I shall not here examine whether the reddish colour of vegetable earth proceeds from the iron which is contained in the earths that are deposited by the rains and dews, but being of importance, shall take notice of it when we come to treat of minerals; it is sufficient to have explained our conception of the formation of the superficial strata of the earth, and by other examples we shall prove, that the formation of the interior strata, can only be the work of the waters.

The surface of the globe, says Woodward, this external stratum on which men and animals walk, which serves as a magazine for the formation of vegetables and animals, is, for the greatest part, composed of vegetable or animal matter, and is in continual motion and variation. All animals and vegetables which have existed from the creation of the world, have successively extracted from this stratum the matter which composes it, and have, after their deaths, restored to it this borrowed matter: it remains there always ready to be retaken, and to serve for the formation of other bodies of the same species successively, for the matter which composes one body is proper and natural to form another body of the same kind. In uninhabited countries, where the woods are never cut, where animals do not brouze on the plants, this stratum of vegetable earth increases considerably. In all woods, even in those which are sometimes cut, there is a bed of mould, of six or eight inches thick, formed entirely by the leaves, small branches, and barks which have perished. I have often observed on the ancient Roman way, which crosses Burgundy in a long extent of soil, that there is formed a bed of black earth more than a foot thick upon the stones, which nourishes very high trees; and this stratum could be composed only of a black mould formed by the leaves, bark, and perished wood. As vegetables inhale for their nutriment much more from the air and water than the earth, it happens that when they perish, they return to the earth more than they have taken from it. Besides, forests collect the rain water, and by stopping the vapours increase their moisture; so in a wood which is preserved a long time, the stratum of earth which serves for vegetation increases considerably. But animals restoring less to the earth than they take from it, and men making enormous consumption of wood and plants for fire, and other uses, it follows that the vegetable soil of inhabited countries must diminish, and become, in time, like the soil of Arabia Petrea, and other eastern provinces, which, in fact, are the most ancient inhabited countries, where only sand and salt are now to be met with; for the fixed salts of plants and animals remain, whereas all the other parts volatilise, and are transported by the air.

Let us now examine the position and formation of the interior strata: the earth, says Woodward, appears in places that have been dug, composed of strata placed one on the other, as so many sediments which necessarily fell to the bottom of the water; the deepest strata are generally the thickest, and those above the thinnest, and so gradually lessening to the surface. We find sea shells, teeth, and bones of fish in these different beds, and not only in those that are soft, as chalk and clay, but even in those of hard stone, marble, &c. These marine productions are incorporated with the stone, and when separated from them, leave the impressions of the shells with the greatest exactness. "I have been most clearly and positively assured," says this author, "that in France, Flanders, Holland, Spain, Italy, Germany, Denmark, Norway, and Sweden, stone, and other terrestrial substances are disposed in strata, precisely the same as they are in England; that these strata are divided by parallel fissures; that there are inclosed within stones and other terrestrial and compact substances, a great quantity of shells and other productions of the sea, disposed in the same manner as in this island. I am also informed that these strata are found the same in Barbary, Egypt, Guinea, and in other parts of Africa; in Arabia, Syria, Persia, Malabar, China, and the rest of the provinces of Asia; in Jamaica, Barbadoes, Virginia, New-England, Brazil, and other parts of America[20]."

This author does not say how he learnt, or by whom he was told, that the strata of Peru contained shells; yet as in general his observations are exact, I do not doubt but he was well informed; and am persuaded that shells may be found in the earth of Peru, as well as elsewhere. This remark is made from a doubt having been formed some time since on the subject, and which I shall hereafter consider.

In a trench made at Amsterdam, to the depth of 230 feet, the strata were found as follows: 7 feet of vegetable earth, 9 of turf, 9 of soft clay, 8 of sand, 4 of earth, 10 of clay, 4 of earth, 10 of sand, then 2 feet of clay, 4 of white sand, 5 of dry earth, 1 of soft earth, 14 of sand, 8 of argil, mixed with earth; 4 of sand, mixed with shells; then clay 102 feet thick, and at last 31 feet of sand, at which depth they ceased digging[21].

---

[20] Essay on the Natural History of the Earth, pages 40, 41, 42, &c.
[21] See Varennii, Geograph. General, page 46.

It is very singular to dig so deep without meeting with water: and this circumstance is remarkable in many particulars. 1. It shews, that the water of the sea does not communicate with the interior part of the earth, by means of filtration. 2. That shells are found at the depth of 100 feet below the surface, and that consequently the soil of Holland has been raised 100 feet by the sediment of the sea. 3. We may draw an induction, that this strata of thick clay of 102 feet, and the bed of sand below it, in which they dug to 31 feet, and whose entire thickness is unknown, are perhaps not very far distant from the first strata of the original earth, such as it was before the motion of the water had changed its surface. We have said in the first article, that if we desired to find the ancient earth, we should dig in the northern countries, rather than towards the south; in plains rather than in mountainous regions. The circumstances in this instance, appear to be nearly so, only it is to be wished they had continued the digging to a greater depth, and that the author had informed us, whether there were not shells and other marine productions, in the last bed of clay, and in that of sand below it. The experiment confirms what we have already said; and the more we dig, the greater thickness we shall find the strata.

The earth is composed of parallel and horizontal beds, not only in plains, but hills and mountains are in general composed after the same manner: it may be said, that the strata in hills and mountains are more apparent there than in the plains, because the plains are generally covered with a very considerable quantity of sand and earth, which the water has brought from the higher grounds, and therefore, to find the ancient strata, must dig deeper in the plains than in the mountains.

I have often observed, that when a mountain is level at its summit, the strata which compose it are also level; but if the summit is not placed horizontally, the strata inclines also in the same direction. I have heard that, in general, the beds of quarries inclined a little to the east; but having myself observed all the chains of rocks which offered, I discovered this opinion to be erroneous, and that the strata inclines to the same side as the hill, whether it be east, west, north, or south. When we dig stone and marble from the quarry, we take great care to separate them according to their natural position, and we cannot even get them of a large size, if we cut them in any other direction. Where they are made use of for good masonry, the workmen are particular in placing them as they stood in the quarry, for if they were placed in any other direction, they would split, and would not resist the weight with which they are loaded. This perfectly confirms that stones, are found in parallel and horizontal strata, which have been successively heaped one on the other, and that these strata composed masses where resistance is greater in that direction than in any other.

Every strata, whether horizontal or inclined, has an equal thickness throughout its whole extent. In the quarries about Paris the bed of good stone is not thick, scarcely more than 18 or 20 feet: in those of Burgundy the stone is much thicker. It is the same with marble; the black and white marble have a thicker bed than the coloured; and I know beds of very hard stone, which the farmers in Burgundy make use of to cover their houses, that are not above an inch thick. The different strata vary much in thickness, but each bed preserves the same thickness throughout its extent. The thickness of strata is so greatly varied, that it is found from less than a line to 1, 10, 20, 30, or 100 feet thick. The ancient and modern quarries, which are horizontally dug, the perpendicular and other divisions of mines, prove that there are extensive strata in all directions. "It is thoroughly proved," says the historian of the academy, "that all stones have formerly been a soft paste, and as there are quarries almost in every part, the surface of the earth has therefore consisted, in all these places, of mud and slime, at least to certain depths. The shells found in most quarries prove that this mud was an earth diluted by the water of the sea, and consequently that the sea covered all these places; and it could not cover them without also covering all that was level with or lower than it: and it is plain that it could not cover every place where there were quarries, without covering the whole face of the terrestrial globe. We do not here consider the mountains which the sea must also at one time have covered, since quarries and shells are often found in them.

"The sea," continues he, "therefore, covered the whole earth, and from thence it proceeds that all the beds of stone in the plains are horizontal and parallel; fish must have also been the most ancient inhabitants of the globe, as there was no sustenance for either birds or terrestrial animals." But how did the sea retire into these vast basins which it at present occupies? What presents itself the most natural to the mind is, that the earth, at least at a certain depth, was not entirely solid, but intermixed with some great vacuums, whose vaults were supported for a time, but at length, sunk in suddenly: then the waters must have fallen into these vacancies, filled them, and left naked a part of the earth's surface, which became an agreeable abode to terrestrial animals and birds. The shells found in quarries perfectly agree with this idea, for only the bony parts of fish could be preserved till now. In general, shells are heaped up in great abundance in certain parts of the sea, where they are immovable, and form a kind of rock, and could not follow the water, which suddenly forsook them: this is the reason that we find more shells than bones of the fish, and this even proves a sudden fall of the sea into its present basins. At the same time as our supposed vaults gave way, it is very possible that other parts of the globe were raised by the same cause, and that mountains were placed on this surface with quarries already formed, but the beds of these quarries could not preserve the horizontal direction they before had, unless the mountains were raised precisely perpendicular to the surface of the earth, which could happen but very seldom: so also, as we have already observed, in 1705, the beds of stone in mountains are always inclined to the horizon, though parallel with each other; for they have not changed their position with respect to each other, but only with respect to the surface of the earth[22].

These parallel strata, these beds of earth and stone, which have been formed by the sediment of the sea, often extend to considerable distances, and we often find in hills, separated by a valley, the same beds and the same matters at the same level. This observation agrees perfectly with that of the height of the opposite hills. We may easily be assured of the truth of these facts, for in all narrow vallies, where rocks are discovered, we shall find the same beds of stone and marble on both

---

[22] See the Mem. of the Acad. 1716, page 14.

sides at the same height. In a country where I frequently reside, I found a quarry of marble which extended more than 12 leagues in length, and whose breadth was very considerable, although I have never been able precisely to determine it. I have often observed that this bed of marble is throughout of the same thickness, and in hills divided from this quarry by a valley of 100 feet depth, and a quarter of a mile in breadth, I found the same bed of marble at the same height. I am persuaded it is the same in every stone and marble quarry where shells are found; but this observation does not hold good in quarries of freestone. In the course of this work, we shall give reasons for this difference, and describe why freestone is not dispersed, like other matters, in horizontal beds, and why it is in irregular blocks, both in form and position.

We have likewise observed that the strata are the same on both sides the straits of the sea. This observation, which is important, may lead us to discover the lands and islands which have been separated from the continent; it proves, for example, that England has been divided from France; Spain from Africa; Sicily from Italy; and it is to be wished that the same observation had been made in all the straits. I am persuaded that we should find it almost every where true. We do not know whether the same beds of stone are found at the same height on both sides the straits of Magellan, which is the longest; but we see, by the particular maps and exact charts, that the two high coasts which confine it, form nearly, like the mountains of the earth, correspondent angles, which also proves that the Terra del Fuega, must be regarded as part of the continent of America; it is the same with Forbisher's Strait and the island of Friesland, which appear to have been divided from the continent of Greenland.

The Maldivian islands are only separated by small tracts of the sea, on each side of which banks and rocks are found composed of the same materials; and these islands, which, taken together, are near 200 miles long, formed anciently only one land; they are now divided into 13 provinces, called Clusters. Each cluster contains a great number of small islands, most of which are sometimes overflowed and sometimes dry; but what is remarkable, these thirteen clusters are each surrounded with a chain of rocks of the same stone, and there are only three or four dangerous inlets by which they can be entered. They are all placed one after the other, and it evidently appears that these islands were formerly a long mountain capped with rocks[23].

Many authors, as Verstegan, Twine, Somner, and especially Campbell, in his Description of England, in the chapter of Kent, gives very strong reasons, to prove that England was formerly joined to France, and has been separated from it by an effort of the sea, which carried away the neck of land that joined them, opened the channel, and left naked a great quantity of low and marshy ground along the southern coasts of England. Dr. Wallis, as a corroboration of this supposition, shews the conformity of the ancient Gallic and British tongues, and adds many observations, which we shall relate in the following articles.

If we consider the form of lands, the position of mountains, and the windings of rivers, we shall perceive that generally opposite hills are not only composed of

[23] See the Voyages of Francis Piriard, vol. 1, page 108.

the same matters on the same level, but are nearly of an equal height. This equality I have observed in my travels, and have mostly found them the same on the two sides, especially in vallies that were not more than a quarter or a third of a league broad, for in vallies which are very broad, it is difficult to judge of the height and equality of hills, because, by looking over a level plain of any great extent, it appears to rise, and hills at a distance appear to lower; but this is not the place to give a mathematical reason for this difference. It is also very difficult to judge by the naked sight of the middle of a great valley, at least if there is no river in it; whereas in confined vallies our sight is less equivocal and our judgment more certain. That part of Burgundy comprehended between Auxerre, Dijon, Autun, and Bar-sur-seine, a considerable extent of which is called *la Bailliage de la Montagne*, is one of the highest parts of France; from one side of most of these mountains, which are only of the second class, the water flows towards the Ocean, and on the other side towards the Mediterranean. This high country is divided with many small vallies, very confined, and almost all watered with rivulets. I have a thousand times observed the correspondence of the angles of these hills and their equality of height, and I am certain that I have every where found the saliant angles opposite to the returning angles, and the heights nearly equal on both sides. The farther we advance into the higher country, where the points of division are, the higher are the mountains; but this height is always the same on both sides of the vallies, and the hills are raised or lowered alike. I have frequently made the like observations in many other parts of France. It is this equality in the height of the hills which forms the plains in the mountains, and these plains form lands higher than others. But high mountains do not appear so equal in height, most of them terminate in points and irregular peaks; and I have seen, in crossing the Alps, and the Apennine mountains, that the angles are, in fact, correspondent; but it is almost impossible to judge by the eye of the equality or inequality in the height of opposite mountains, because their summits are lost in mists and clouds.

The different strata of which the earth is composed are not disposed according to their specific weight, for we often find strata of heavy matters placed on those of lighter. To be assured of this, we have only to examine the earth on which rocks are placed, and we shall find that it is generally clay or sand, which is specifically lighter. In hills, and other small elevations, we easily discover this to be the case; but it is not so with large mountains, for not only their summits are rocks, but those rocks are placed on others; there mountains are placed upon mountains, and rocks upon rocks, to such a considerable height, and through so great an extent of country, that we can scarcely be certain whether there is earth at bottom, or of what nature it is. I have seen cavities made in rocks to some hundred feet deep, without being able to form an idea where they ended, for these rocks were supported by others; nevertheless, may we not compare great with small? and since the rocks of little mountains, whose bases are to be seen, rest on the earth less heavy and solid than stone, may we not suppose that earth is also the base of high mountains? All that I have here to prove by these arguments is, that, by the motion of the waters, it may naturally happen that the more ponderous matters accumulated on the lighter; and that, if this in fact is found to be so in most hills, it is probable that it happened as explained by my theory; but should it be objected that I am not grounded in supposing, that before the formation of mountains the heaviest matters were below the lighter; I answer, that I assert nothing general in this respect, because this effect may have been produced in many manners, whether the heaviest matters were uppermost or undermost, or placed indiscriminately. To conceive how the sea at first formed a mountain of clay, and afterwards capt it with rocks, it is sufficient to consider the sediments may successively come from different parts, and that they might be of different materials. In some parts, the sea may at first have deposited sediments of clay, and the waters afterwards brought sediment of strong matter, either because they had transported all the clay from the bottom and sides, and then the waves attacked the rocks, possibly because the first sediment came from one part, and the second from another. This perfectly agrees with observation, by which we perceive that beds of earth, stone, gravel, sand, &c. followed no rule in their arrangement, but are placed indifferently one on the other as it were by chance.

But this chance must have some rules, which can be known only by estimating the value of probabilities, and the truth of conjectures. According to our hypothesis, on the formation of the globe, we have seen that the interior part of the globe must have been a vitrified matter, similar to vitrified sand, which is only the fragments of glass, and of which the clays are perhaps the scoria; by this supposition, the centre of the earth, and almost as far as the external circumference, must be glass, or a vitrified matter; and above this we shall find sand, clay, and other scoria. Thus the earth, in its first state, was a nucleus of glass, or vitrified matter; either massive like glass, or divided like sand, because that depends on the degree of heat it has undergone. Above this matter was sand, and lastly clay. The soil of the waters and air produced the external crust, which is thicker or thinner, according to the situation of the ground; more or less coloured, according to the different mixtures of mud, sand, clay, and the decayed parts of animals and vegetables; and more or less fertile, according to the abundance or want of these parts. To shew that this supposition on the formation of sand and clay is not chimerical, I shall add some particular remarks.

I conceive, that the earth, in its first state, was a globe, or rather a spheroid of compact glass, covered with a light crust of pumice stone and other scoria of the matter in fusion. The motion and agitation of the waters and air soon reduced this crust into powder or sand, which, by uniting afterwards, produced flints, and owe their hardness, colour, or transparency and variety, to the different degrees of purity of the sand which entered into their composition.

These sands, whose constituting parts unite by fire, assimilate, and become very dense, compact, and the more transparent as the sand is more pure; on the contrary, being exposed a long time to the air, they disunite and exfoliate, descend in the form of earth, and it is probable the different clays are thus produced. This dust, sometimes of a brightish yellow, and sometimes like silver, is nothing else but a very pure sand somewhat perished, and almost reduced to an elementary state. By time, particles will be so far attenuated and divided, that they will no longer have power to reflect the light, and acquire all the properties of clay.

This theory is conformable to what every day is seen; let us immediately wash sand upon its being dug, and the water will be loaded with a black ductile and fat earth, which is genuine clay. In streets paved with freestone, the dirt is always black and greasy, and when dried appears to be an earth of the same nature as clay. Let us wash the earth taken from a spot where there are neither freestone nor flints, and there will always precipitate a great quantity of vitrifiable sand.

But what perfectly proves that sand, and even flint and glass, exist in clay, is, that the action of fire, by uniting the parts, restores it to its original form. Clay, if heated to the degree of calcination, will cover itself with a very hard enamel; if it is not vitrified internally, it nevertheless will have acquired a very great hardness, so as to resist the file; it will emit fire under the hammer, and it has all the properties of flint; a greater degree of heat causes it to flow, and converts it into real glass.

Clay and sand are therefore matters perfectly analogous, and of the same class; if clay, by condensing, may become flint and glass, why may not sand, by dissolution, become clay? Glass appears to be true elementary earth, and all mixed substances disguised glass. Metals, minerals, salts, &c. are only vitrifiable earth; common stone and other matters analogous to it, and testaceous and crustaceous shells, &c. are the only substances which cannot be vitrified, and which seem to form a separate class. Fire, by uniting the divided parts of the first, forms an homogeneous matter, hard and transparent, without any diminution of weight, and to which it is not possible to cause any alteration; those, on the contrary, in which a greater quantity of active and volatile principles enter, and which calcine, lose more than one-third of their weight in the fire, and retake the form of simple earth, without any other alteration than a disunion of their different parts: these bodies excepted, which are no great number, and whose combinations produce no great varieties in nature, every other substance, and particularly clay, may be converted into glass, and are consequently only decomposed glass. If the fire suddenly causes the form of these substances to change, by vitrifying them, glass itself, whether pure, or in the form of sand or flint, naturally, but by a slow and insensible progress, changes into clay.

Where flint is the predominant stone, the country is generally strewed with parts of it, and if the place is uncultivated, and these stones have been long exposed to the air, without having been stirred, their upper superficies is always white, whereas the opposite side, which touches the earth, is very brown, and preserves its natural colour. If these flints are broken, we shall perceive that the whiteness is not only external, but penetrates internally, and there forms a kind of band, not very deep in some, but which in others occupies almost the whole flint. This white part is somewhat grainy, entirely opaque, as soft as freestone, and adheres to the tongue like the boles; whereas the other part is smooth, has neither thread nor grain, and preserves its natural colour, transparency, and hardness. If this flint is put into a furnace, its white part becomes of a brick colour, and its brown part of a very fine white. Let us not say with one of our most celebrated naturalists, that these stones are imperfect flints of different ages, which have not acquired their perfection; for why should they be all imperfect? Why should they be imperfect only on the side exposed to the weather? It, on the contrary, appears to me more reasonable that they are flints changed from their original state, gradually decomposed, and assuming the form and property of clay or bole. If this is thought to be only conjecture, let the hardest and blackest flint be exposed to the weather, in less than a year its surface will change colour; and if we have patience to pursue this experiment, we shall see it by degrees lose its hardness, transparency, and other specific characters, and approach every day nearer and nearer the nature of clay.

What happens to flint happens to sand; each grain of sand may possibly be considered as a small flint, and each flint as a mass of extremely fine grains of sand. The first example of the decomposition of sand is found in the brilliant opaque powder called Mica, in which clay and slate are always diffused. The entirely transparent flints, the Quartz, produce, by decomposition, fat and soft talks, such as those of Venice and Russia, which are as ductile and vitrifiable as clay: and it appears to me, that talk is a mediate between glass, or transparent flint, and clay; whereas coarse and impure flint, by decomposing, passes to clay without any intermedium.

Our factitious glass undergoes the same alterations: it decomposes and perishes, as it were, in the air. At first, it assumes a variety of colours, then exfoliates, and by working it, we perceive brilliant scales fall off; but when its decomposition is more advanced, it crumbles between the fingers, and is reduced into a very white fine talky powder. Art has even imitated nature in the decomposition of glass and flint. "Est etiam certa methodus solius aquæ communis ope, silices & arenam in liquorem viscosum, eumdemque in sal viride convertendi, & hoc in aleum rubicundum, &c. Solius ignis & aqua ope, speciali experimento, durissimos quosque lapides in mucorem resolvo, qui distillan subtilem spiritum exhibet & oleum nullus laudibus prœdicabile[24]."

These matters more particularly belong to metals, and when we come to them, shall be fully treated on, therefore we shall content ourselves here with adding, that the different strata which cover the terrestrial globe, being materials to be considered as actual vitrifications or analogous to glass, and possessing its most essential qualities; and as it is evident, that from the decomposition of glass and flint, which is every day made before our eyes, a genuine clay remains, it is not a precarious supposition to advance, that clays and sands have been formed by scoria, and vitrified drops of the terrestrial globe, especially when we join the proofs *a priori*, which we have given to evince the earth has been in a state of liquefaction caused by fire.

---

[24] See Becher. Phys. subter.

ARTICLE VIII.

ON SHELLS, AND OTHER MARINE PRODUCTIONS FOUND IN THE INTERIOR PARTS OF THE EARTH.

I have often examined quarries, the banks of which were filled with shells; I have seen entire hills composed of them, and chains of rocks which contained them throughout their whole extent. The quantity of these marine productions is astonishing, and the number in many places so prodigious, that it appears scarcely possible that any should now remain in the sea; it is by considering this innumerable multitude of shells, that no doubt is left of our earth having been a long time under the water of the ocean. The quantity found in a fossil, or petrified state, is beyond conception, and it is only from the number of those that have been discovered that we could possibly have formed an idea of their multiplicity. We must imagine, like those who reason on matters they never saw, that shells are only found at random, dispersed here and there, or in small heaps, as oyster shells thrown before our doors; on the contrary, they form mountains, are met with in shoals of 100 or 200 miles length, nay, they may sometimes be traced through whole provinces in masses of 50 or 60 feet thick. It is from these circumstances alone that we can reason on the subject.

We cannot give a more striking example on this subject than the shells of Touraine. The following is the description given of them by the historian of the Academy[25].

"The number of figured stones and fossil shells found in the bowels of the earth were remarked in all ages and nations, but they were considered merely as the sports of nature, and even by philosophers themselves, as the productions of chance or accident; they regarded them with a degree of surprise, but passed them over with a slight attention, and all this phenomena perished without any fruit for the progress of knowledge. A potter in Paris, who knew neither Latin nor Greek, towards the end of the 16th century, was the first man who dared affirm, in opposition to the learned, that the fossil shells were real shells formerly deposited by the sea in those places where they were found; that animals, and particularly fish, had given to stones all these different figures, &c. and he desired the whole school of Aristotle to contradict his proofs. This was Bernard Palissy, as great a natural genius as nature could form: his system slept near 100 years, and even his name was almost forgot. At length the ideas of Palissy were revived in the mind of several philosophers; and science has profited by all the shells and figured stones

---

[25] Anno 1720; page 5.

the earth furnishes us with; perhaps they are at present become only too common, and the consequences drawn from them too incontestable.

"Notwithstanding this, the observations presented by M. Reaumer must appear wonderful. He discovered a mass of 130 million, 680 thousand cubical fathoms of shells, either whole or in fragments, without any mixture of stone, earth, sand, or other extraneous matter: hitherto fossil shells have never appeared in such an enormous quantity, nor without mixture. It is in Touraine this prodigious mass is found, more than 36 leagues from the sea; this is perfectly known there, as the farmers of that province make use of these shells, which they dig up, as manure for their lands, to fertilize their plains, which otherwise would be absolutely sterile.

"What is dug from the earth, and which generally is no more than eight or nine feet deep, are only small fragments of shells, very distinguishable as fragments, for they retain their original channels and hollows, having only lost their gloss and colour, as almost all shells do which we find in the earth. The smallest pieces, which are only dust, are still distinguishable because they are perfectly of the same matter as the rest, as well as of the whole shells which are sometimes found. We discover the species as well in the whole shells as in the larger fragments. Some of these species are known at Poictou, others belong to more remote coasts. There are even fragments of madrepores, coral, and other productions of the sea; all this matter in the country is termed *Fallun*, and is found wherever the ground is dug in that province for the space of nine leagues square. The peasants do not dig above twenty feet deep, because they think it would not repay them for their trouble, but they are certainly deeper. The calculation of the quantity is however taken upon the supposition of only 18 feet and 2200 fathoms to the league. This mass of shells of course exceeds the calculation, and possibly contains double the quantity.

"In physical points the smallest circumstances, which most people do not think worthy of remarking, sometimes lead to consequences and afford great lights. M. de Reaumer observed, that all these fragments of shells lie horizontally, and hence he has concluded that this infinity of fragments does not proceed from the heap being formed at one time, or of whole shells, for the uppermost, by their weight, would have crushed the others, and of course their fallings would have given an infinity of different positions. They must, therefore, have been brought there by the sea, either whole or broken, and necessarily placed horizontal; and although the extreme length of time was of itself sufficient to break, and almost calcine the greatest part, it could not change their position.

"By this it appears, that they must have been brought gradually, and, in fact, how was it possible that the sea could convey at once such an immense quantity of shells, and at the same time preserve a position perfectly horizontal? they must have collected in one spot, and consequently this spot must have been the bottom of a gulph or basin.

"All this proves, that although there must remain upon the earth many vestiges of the universal deluge, as recorded in scripture, the mass of shells at Touraine was not produced by that deluge; there is perhaps not so great a mass in any part of the sea; but even had the deluge forced them away, it would have been with an impetuosity and violence that would not have permitted them to retain one uniform position. They must have been brought and deposited gently and slowly, and consequently their accumulation required a space of time much longer than a year."

The surface of the earth, it is evident, must have been before or after the deluge very differently disposed to what it is at present, that the sea and continent had another arrangement, and formerly there was a great gulph in the middle of Touraine. The changes which are known from history, or even ancient fable, are inconsiderable, but they give us room to imagine those which a longer time might bring about. M. de Reaumur supposes that Touraine was a gulph of the sea which communicated with the ocean, and that the shells were carried there by a current; but this is a simple conjecture laid down in room of the real unknown fact. To speak with certainty on this matter, we should have geographical maps of all the places where shells have been dug from the earth, to obtain which would require almost an infinity of time and observation, yet it is possible that hereafter science may accomplish it.

This quantity of shells, considerable as it is, will astonish us less if we consider the following circumstances: first, shell fish multiply prodigiously, and are full grown in a very short time; the abundance of individuals in each kind proves to us their fertility. We have a strong example of this increase in oysters, a mass of many fathoms of which are frequently raised in a single day. In a very short time the rocks to which they are attached are considerably diminished, and some banks quite exhausted, nevertheless the ensuing year we find them as plentiful as before, nor do they appear to be in the least diminished; indeed I know not whether a natural bed of oysters was ever entirely exhausted. Secondly, the substance of shells is analogous to stone; they are a long time preserved in soft matters, and petrify readily in hard; these shells and marine productions therefore found on the earth, being the wrecks of many ages, must of course have formed very considerable masses.

There are a prodigious quantity of shells in marble, lime, stone, chalk, marl, &c. we find them, as before observed, in hills and mountains, and they often make more than one half of the bodies which contain them; for the most part they appear well preserved, others are in fragments, but large enough to distinguish to what kind of shells they belong. Here our knowledge on this subject, from observation, finds its limits; but I shall go further and assert that shells are the intermedium which Nature adopts for the formation of most kind of stones; that chalks, marls, and lime-stone are composed only of the powder and pieces of shells; that consequently the quantities of shells destroyed are infinitely more considerable than those preserved. I shall here content myself with indicating the point of view in which we ought to consider the strata of which the globe is composed. The first stratum is composed of the dust of the air, the sediment of the rain, dew, and vegetable or animal parts, reduced to particles; the strata of chalk, marl, lime, stone, and marble, are composed of the ruins of shells, and other marine productions, mixed with fragments or whole shells; but the vitrifiable sand or clay are the matters of which the internal parts of the globe are composed. They were vitrified when the globe received its form, which necessarily supposes that the matter was in fusion. The granate, rock, flint, &c. owe their origin to sand and clay, and are likewise disposed by strata; but tuffa[26], free-stone, and flints (not in great masses), crystals, metals, pyrites, most minerals, sulphurs, &c. are matters whose formation is novel, in comparison with marbles, calcinable stones, chalk, marl, and all other materials disposed in horizontal strata, and which contain shells and other productions of the sea.

As the denominations I make use of may appear obscure or equivocal, it is necessary to explain them. By the term *clay*, I mean not only the white and yellow, but also blue, soft, hard, foliated, and other clays, which I look on as the scoria of glass, or as decomposed glass. By the word *sand* I always understand vitrifiable sand; and not only comprehend under this denomination the fine sand which produces freestone, and which I look upon as powdered glass, or rather pumice stone, but also the sand which proceeds from the freestone destroyed by friction, and also the larger sand, as small gravel, which proceeds from the granate and rock-stone, and is sharp, angular, red, and commonly found in the bed of rivers or rivulets that derive their waters immediately from the higher mountains, or hills composed of stone or granate. The river Armanson conveys a great quantity of this sand; it is large and brittle, and in fact is only fragments of rock-stone, as calcinable gravel is of freestone. *Rock-stone* and *granate* are one and the same substance, but I have used both denominations, because there are many persons who make two different species of them. It is the same with respect to flints and free-stone in large pieces; I look on them as kinds of granate, and I call them *large flints*, because they are disposed like calcinable stone in strata, and to distinguish them from the flints and free-stone in small masses, and the round flints which have no regular quarries, and whose beds have a certain extent; these are of a modern formation, and have not the same origin as the flints and free-stone in large lumps, which are disposed in regular strata.

I understand by the term *slate*, not only the blue, which all the world knows, but white, grey, and red slate: these bodies are generally met with below laminated clay, and have every appearance of being nothing more than clay hardened in this strata. Pit coal and jet are matters which also belong to clay, and are commonly under slate. By the word *tuffa*, I understood not only the common pumice which appears full of holes, and, as I may say, organized, but all the beds of stone made by the sediment of running waters, all the stalactites, incrustations, and all kinds of stone that dissolve by fire. It is no ways doubtful that these matters are not modern, and that they every day grow. Tuffa is only a mass of lapidific matter in which we perceive no distinct strata: this matter is disposed generally in small hollow cylinders, irregularly grouped and formed by waters dropt at the foot of mountains, or on the slope of hills, which contain beds of marl or soft and calcareous earth; these cylinders, which make one of the specific characters of this kind of tuffa, is either oblique or vertical according to the direction of the streams or water which form them. These sort of spurious quarries have no continuation; their extent is very confined, and proportionate to the height of the mountains which furnish

---

[26] A kind of soft gravelly stone.

them with the matter of their growth. The tuffa every day receiving lapidific juices, those small cylindrical columns, between which intervals are left, close at last, and the whole becomes one compact body, but never acquires the hardness of stone, and is what Agricola terms *Marga tofocea fistulosa*. In this tuffa are generally found impressions of leaves, trees, and plants, like those which grow in the environs: terrestrial shells also are often met with, but never any of the marine kind. The tuffa is certainly therefore a new matter, which must be ranked with stalactites, incrustations, &c. all these new matters are kinds of spurious stones, formed at the expence of the rest, but which never arrive at true petrification.

Crystal, precious stones, and all those which have a regular figure, even small flints formed by concentrical beds, whether found in perpendicular cavities of rocks, or elsewhere, are only exudations of large flints, or concrete juices of the like matters, and are therefore spurious stones, and real stalactites of flint or rock.

Shells are never found either in rock, granate, or free-stone, although they are often met with in vitrifiable sand, from which these matters derive their origin; this seems to prove that sand cannot unite to form free-stone or rock but when it is pure, and that if it is mixed with shells or substances of other kinds, which are heterogeneous to it, its union is prevented. I have observed the little pebbles which are often found in beds of sand mixed with shells, but never found any shell there-in: these pebbles are real concretions of free-stone formed in the sand in the places where it is not mixed with heterogeneous matters which oppose the formation of larger masses.

We have before observed, that at Amsterdam, which is a very low country, sea shells were found at 100 feet below the earth, and at Marly-la-Ville, six miles from Paris, at 75 feet; we likewise meet with the same at the bottom of mines, and in banks of rocks, beneath a height of stone 50, 100, 200, and 1000 feet thick, as is apparent in the Alps and Pyrennees, where, in the lower beds, shells and other marine productions are constantly found. But to proceed in order, we find shells on the mountains of Spain, France, and England; in all the marble quarries of Flanders, in the mountains of Gueldres, in all hills around Paris, Burgundy, and Champagne; in one word, in every place where the basis of the soil is not free-stone or tuffa; and in most of these places there are more shells than other matters in the substance of the stones. By *shells*, I mean not only the wrecks of shell-fish, but those of crustaceous animals, the bristles of sea hedge-hogs, and all productions of the sea insects, as coral, madrepores, astroites, &c. We may easily be convinced by inspection, that in most calculable stones and marble, there is so great a quantity of these marine productions that they appear to surpass the matter which unites them.

But let us proceed; we meet with these marine productions even on the tops of the highest mountains; for example, on Mount Cenis, in the mountains of Genes, in the Apennines, and in most of the stone and marble quarries in Italy; also in the stones of the most ancient edifices of the Romans; in the mountains of Tirol; in the centre of Italy, on the summits of Mount Paterne, near Bologna; in the hills of Calabria; in many parts of Germany and Hungary, and generally in all the high parts of Europe[27].

In Asia and Africa, travellers have remarked them in several parts; for example, on the mountains of Castravan, above Barut, there is a bed of white stone as thin as slate, each leaf of which contains a great number and diversity of fishes; they lie for the most part very flat and compressed, as does the fossil fearn-plants, but they are notwithstanding so well preserved, that the smallest traces of the fins, scales, and all the parts which distinguish each kind of fish, are perfectly visible. So likewise we find many sea muscles, and petrified shells between Suez and Cairo, and on all the hills and eminences of Barbary; the greatest part are conformable to the kinds at present caught in the Red Sea[28]. In Europe, we meet with petrified fish in Sweden and Germany, and in the quarry of Oningen, &c.

The long chain of mountains, says Bourguet, which extends from Portugal to the most eastern parts of China, the mountains of Africa and America, and the vallies of Europe, all inclose stones filled with shell-fish, and from hence, he says, we may conclude the same of all the other parts of the world unknown to us.

The islands in Europe, Asia, and America, where men have had occasion to dig, whether in mountains or plains, furnish examples of fossil shells, which evince that they have that in common with the bordering continents.

Here then is sufficient facts to prove that sea shells, petrified fish, and other marine productions are to be found in almost every place we are disposed to seek them.

"It is certain, says an English author (Tancred Robinson), that there have been sea-shells dispersed on the earth by armies, and the inhabitants of towns and villages, and that Loubere relates in his Voyage to Siam, that the monkies of the Cape of Good Hope, continually amuse themselves with carrying shells from the sea shores to the tops of the mountains; but that cannot resolve the question, why these shells are dispersed over all the earth, and even in the interior parts of mountains, where they are deposited in beds like those in the bottom of the sea."

On reading an Italian letter on the changes happened to the terrestrial globe, printed at Paris in the year 1746, I was surprised to find these sentiments of Loubere exactly corresponded. Petrified fish, according to this writer, are only fish rejected from the Roman tables, because they were not esteemed wholesome; and with respect to fossil-shells, he says the pilgrims of Syria brought, during the times of the Crusades, those of the Levant Sea, into France, Italy, and other Christian states; why has he not added that it was the monkies who transported the shells to the tops of these mountains, which were never inhabited by men? This would not have spoiled but rendered his explanation still more probable.

How comes it that enlightened persons, who pique themselves on philosophy, have such various ideas on this subject? But doing so, we shall not content ourselves with having said that petrified shells are found in almost every part of the earth which has been dug, nor with having related the testimonies of authors of natural history; as it might be suspected, that with a view of some system, they

[27] On this subject see Stenon, Ray, Woodward, and others.
[28] See Shaw's Voyages, Vol. ii, pages 40 and 41.

perceived shells where there were none; but quote the authority of some authors, who merely remarked them accidentally, and whose observations went no farther than recognising those that were whole and in the best preservation. Their testimony will perhaps be of a still greater authority with people who have it not in their power to be assured of the truth of these facts, and who know not the difference between shells and petrifications.

All the world may see the banks of shells in the hills in the environs of Pans, especially in the quarries of stone, as at Chaussée, near Séve, at Issy, Passy, and elsewhere. We find a great quantity of lenticular stones at Villers-Cotterets; these rocks are entirely formed thereof, and they are blended without any order with a kind of stony mortar, which binds them together. At Chaumont so great a quantity of petrified shells are found that the hills appear to be composed of nothing else. It is the same at Courtagnon, near Rheims, where there is a bank of shells near four leagues broad, and whose length is considerably more. I mention these places as being famous and striking the eye of every beholder.

With respect to foreign countries, here follows the observations of some travellers:

"In Syria and Phœnicia, the rocks, particularly in the neighbourhood of Latikea, are a kind of chalky substance, and it is perhaps from thence that the city has taken the name of the white promontory. Nakoura, anciently termed Scala Tyriorum, or the Tyrians Ladder, is nearly of the same nature, and we still find there, by digging, quantities of all sorts of shells, corals, and other remains of the deluge[29]."

On mount Sinai, we find only a few fossil shells, and other marks of the deluge, at least if we do not rank the fossil Tarmarin of the neighbouring mountains of Siam among this number, perhaps the first matter of which their marble is formed, had a corrosive virtue not proper to preserve them. But at Corondel, where the rocks approach nearer our free-stone, I found many shells, as also a very singular sea muscle, of the descoid kind, but closer and rounder. The ruins of the little village Ain le Mousa, and many canals which conduct the water thereto, furnish numbers of fossil shells. The ancient walls of Suez, and what yet remains of its harbour, have been constructed of the same materials, which seem to have been taken from the same quarry. Between, as well as on all the mountains, eminences and hills of Lybia, near Egypt, we meet with a great quantity of sea weed, as well as vivalvous shells, and of these which terminate in a point, most of which are exactly conformable to the kinds at present caught in the Red Sea.

The moving sand in the neighbourhood of Ras Sem, in the kingdom of Barca, covers many palm trees with petrifications. Ras Sem signifies the head of a fish, and is what we term the petrified village, where it is said men, women, and children are found, who with their cattle, furniture, &c. have been converted into stone; but these, says Shaw, are vain tales and fables, as I have not only learnt from M. le Maire, who at the time he was Consul at Tripoly, sent several persons thither to take cognizance of it, but also from very respectable persons who had been at those places.

---

[29] See Shaw's Travels.

Near the pyramids certain pieces of stone worked by the sculptor, were found by Mr. Shaw, and among these stones many rude ones of the figure and size of lentils; some even resemble barley half-peeled; these, he says, were reported to be the remains of what the workmen ate, but which does not appear probable, &c. These lentils and barley are nothing but petrified shells called by naturalists lentil stones.

According to Misson, several sorts of these shell-fish are found in the environs of Maestricht, especially towards the village of Zicken, or Tichen, and at the little mountain called Huns. In the environs of Sienna, near Ceraldo, are many mountains of sand crammed with divers sorts of shells. Montemario, a mile from Rome, is entirely filled with them; I have seen them in the Alps, France, and elsewhere. Olearius, Steno, Cambden, Speed, and a number of other authors, as well ancient as modern, relate the same phenomena.

"The island of Cerigo, says Thevenot, was anciently called Porphyris, from the quantity of Porphyry which was taken out of it[30].

"Opposite the village of Inchene, and on the eastern shore of the Nile, I found petrified plants, which grow naturally in a space about two leagues long, by a very moderate breadth; this is one of the most singular productions of nature. These plants resemble the white coral found in the Red Sea[31]."

"There are petrifications of divers kinds on Mount Libanus, and among others flat stones, where the skeletons of fish are found well preserved and entire; red chesnuts and small branches of coral, the same as grow in the Red Sea, are also found on this mountain."

"On Mount Carmel we find a great quantity of hollow stones, which have something of the figure of melons, peaches, and other fruits, which are said to be so petrified: they are commonly sold to pilgrims, not only as mere curiosities, but also as remedies against many disorders. The olives which are the *lapides jadaici*, are to be met with at the druggists, and have always been looked upon, when dissolved in the juice of lemon, as a specific for the stone and gravel."

"M. la Roche, a physician, gave me some of these petrified olives, which grew in great plenty in these mountains, where I am told are found other stones, the inside of which perfectly resemble the natural parts of men and women. These are Hysterolithes."

"In going from Smyrna to Tauris, when we were at Tocat, says Tavernier, the heat was so great, as obliged us to quit the common road, and go by the mountains, where there is constantly shade and refreshing air. In many places we found snow and a quantity of very fine sorrel, and on the top of some of those mountains we found shells like those upon the sea shores, which was very extraordinary."

Here follows what Olearius says on the subject of petrified shells, which he remarked in Persia, and in the rocks where the sepulchres are cut out, near to the village of Pyrmaraus:

"We were three in company that ascended to the top of the rock by the most

[30] Thevenot, Vol. I, page 25.
[31] Voyage of Paul Lucus, Vol. II, page 380.

frightful precipices, mutually assisting each other; having gained the summit, we found four large chambers, and within many niches cut in the rocks to serve for beds: but what the most surprised us was to find in this vault, on the top of the mountain, muscle shells; and in some places they were in such great quantities, that the whole rock appeared to be composed only of sand and shells. Returning to Persia, we perceived many of these shelly mountains along the coast of the Caspian sea."

To these I could subjoin many other authorities which I suppress, not willing to tire those who have no need of superabundant proofs, and who are convinced by their sight, as I have been, of the existence of shells wherever we chuse to seek for them.

In France, we not only find the shells of the French coast, but also such as have never been seen in those seas. Some philosophers assert, that the quantity of these foreign petrified shells is much greater than those of our climate; but I think this opinion unfounded; for, independent of the shell-fish which inhabit the bottom of the sea, and are seldom brought up by the fishermen, and which consequently may be looked on as foreigners, although they exist in our seas, I see, by comparing the petrifactions with the living analagous animals, there are more of those of our coasts than of others: for example, most of the cockles, muscles, oysters, ear-shells, limpets, nautili, stars, tubulites, corals, madrepores, &c. found in so many places, are certainly the productions of our seas; and though a great number appear which are foreign or unknown, the cornu ammonis, the lapides juduica, &c. yet I am convinced, from repeated observations, that the number of these kinds is small in comparison with the shells of our own coasts: besides, what composes the bottom of almost all our marble and lime-stone but madrepores, astroites, and all those other productions which are formed by sea insects, and formerly called marine plants? Shells, however abundant, form only a small part of these productions, many of which originate in our seas, and particularly in the Mediterranean.

The Red sea produces corals, madrepores, and marine plants in the greatest abundance: no part furnishes a greater variety than the port of Tor; in calm weather so great a quantity present themselves, that the bottom of the sea resembles a forest; some of the branched madrepores are eight or ten feet high. In the Mediterranean sea, at Marseilles, near the coasts of Italy and Sicily; in most of the gulphs of the ocean, around islands, on banks, and in all temperate climates, where the sea is but of a moderate depth, they are very common.

M. Peyssonel was the first who discovered that corals, madrepores, &c. owed their origin to animals, and were not plants as had been supposed. The observation of M. Peyssonel was a long time doubted; some naturalists, at first, rejected it with a kind of disdain, nevertheless they have been obliged since to acknowledge its truth, and the whole world is at length satisfied that these formerly supposed marine plants, are nothing but hives or cells formed by insects, in which they live as fish do in their shells. These bodies were, at first, placed in the class of minerals, then passed into that of vegetables, and now remain fixed in that of animals, the genuine operations of which they must ever be considered.

There are shell-fish which live at the bottom of the sea, and which are never cast on the shore; authors call them Pelogiæ, to distinguish them from the others which they call Litterales. It is to be supposed the cornu ammonis, and some other kinds that are only found in a petrified state, belong to the former, and that they were filled with the stony sediment in the very places they are found. There might also have been certain animals, whose species are perished, and of which number this shell-fish might be ranked. The extraordinary fossil bones found in Siberia, Canada, Ireland, and many other places, seem to confirm this conjecture, for no animal has hitherto been discovered to whom such bones could belong, as they are, for the most part, of an enormous size.

These shells, according to Woodward, are met with from the top to the bottom of quarries, pits, and at the bottom of the deepest mines of Hungary. And Mr. Ray assures us, they are found a thousand feet deep in the rocks which border the isle of Calda, and in Pembrokeshire in England.

Shells are not only found in a petrified state, at great depths, and at the tops of the highest mountains, but there are some met with in their natural condition, and which have the gloss, colours, and lightness of sea-shells; and to convince ourselves entirely of this matter, we have only to compare them with those found on the sea shores. A slight examination will prove that these fossil and petrified shells are the same as those of the sea; they are marked with the same articulations and in the glossopetri, and other teeth of fishes, which are sometimes found adhering to the jaw-bone, the teeth of the fish are remarked to be smooth and worn at the extremities, and that they have been made use of when the animals were alive.

Almost every where on land we meet with fossil-shells, and of those of the same kind, some are small, others large, some young, others old; some imperfect, others extremely perfect and we likewise sometimes see the young ones adhering to the old.

The shell-fish called *purpura* has a long tongue, the extremity of which is bony, and so sharp, that it pierces the shells of other fish; by which means it draws nutriment from them. Shells pierced in this manner are frequently found in the earth, which is an incontestible proof that they formerly inclosed living fish, and existed in those parts where there were the Purpura.

The obelisks of St. Peter's at Rome, according to John of Latran, were said to come from the pyramids of Egypt; they are of red granite, which is a kind of rock-stone, and, as we have observed, contains no shells; but the African and Egyptian marble, and the porphyry said to have been cut from the temple of Solomon, and the palaces of the kings of Egypt, and used at Rome in different buildings, are filled with shells. Red porphyry is composed of an infinite number of prickles of the species of echinus, or sea chesnut; they are placed pretty near each other, and form all the small white spots which are in the porphyry. Each of these white spots has a black one in its centre, which is the section of the longitudinal tube of the prickles of the echinus. At Fichen, three leagues from Dijon, in Burgundy, is a red stone perfectly similar in its composition to porphyry, and which differs from it only in hardness, not being more so than marble; it appears almost formed of prickles of the echini, and its beds are of a very great extent. Many beautiful pieces of workmanship have been made of it in this province, and particularly the steps of the pedestal of the equestrian statue of Louis le Grand, at Dijon.

This species of stone is also found at Montbard, in Burgundy, where there is an extensive quarry; it is not so hard as marble, contains more of the echini, and less of the red matter. From this it appears that the ancient porphyry of Egypt differs only from that of Burgundy in the degree of hardness, and the number of the points of the echini.

With respect to what the curious call green porphyry, I rather suppose it to be a granite than a porphyry; it is not composed of spots like the red porphyry, and its substance appears to be similar to that of a common granite. In Tuscany, in the stone with which the ancient walls of Volatera were built, there are a great quantity of shells, and this wall was built 2500 years ago. Most marbles, porphyries, and other stones of the most ancient buildings, contain shells and other wrecks of marine productions, as well as the marble we at present take from the quarry; therefore it cannot be doubted, independent even of the sacred testimony of holy writ, that before the deluge the earth was composed of the same materials at it is at present.

From all these facts it is plain that petrified shells are found in Europe, Asia, Africa, and in every place where the observations have been made; they are also found in America, in the Brasils; for example, in Tucumama, in Terra Magellinica, and in such a great quantity in the Antilles, that directly below the cultivable land, the bottom of which the inhabitants call lime, is nothing but a composition of shells, madrepores, astroites, and other productions of the sea. These facts would have made me think that shells and other petrified marine productions were to be found in the greatest part of the continent of America, and especially in the mountains, as Woodward asserts; but M. Condamine, who lived several years at Peru, has assured me he could not discover any in the Cordeliers, although he had carefully sought for them. This exception would be singular, and the consequences that might be drawn from it would be still more so; but I own that, in spite of the testimony of this celebrated naturalist, I am much inclined to suppose, that in the mountains of Peru, as well as elsewhere, there are shells and other marine petrifications, although they have not been discovered. It is well known, that in matter of testimonies, two positive witnesses, who assert to have seen a thing, is sufficient to make a complete proof; whereas ten thousand negative witnesses, and who can only assert not to have seen a thing, can only raise a slight doubt. This reason, united with the strength of analogy, induces me to persist in thinking the shells will be found on the mountains of Peru, especially if we search for them on the rise of the mountain, and not at the summit.

The tops of the highest mountains are generally composed of rock, stone granite, and other vitrifiable matters, which contain no shells.

All these matters were formed out of the beds of the sand of the sea, which covered the tops of these mountains. When the sea left them, the sand and other light bodies were carried by the waters into the plains, so that there remained only rocks on the tops of the mountains, which had been formed under those beds of sand. At two, three, or four hundred fathoms below the tops of these mountains, are often found marble and other calcinable matter, which are disposed in parallel strata, and contain shells and other marine productions; therefore it is not surprising that M. de la Condamine did not find any shells on these mountains, especially if he sought for them in the elevated parts of those mountains which are composed of rock, free-stone, or vitrifiable sand; but had he examined the lower parts of the Cordeliers, he would undoubtedly have found strata of stone, marble, earth, &c. mixed with shells; for in every country where observations have been made, such beds have always been met with.

But suppose that in fact there are no marine productions in the mountains of Peru, all that may be concluded from it will no ways affect our theory; and it might be possible, that there are some parts of the globe which never were covered with water, especially of such elevation as the Cordeliers. But in this case there might be some curious observations made on those mountains, for they would not be composed of parallel strata, the materials also would be very different from those we are acquainted with; they would not have perpendicular cracks; the composition of the rocks and stones would not at all resemble those of other countries; and lastly, in these mountains we should find the ancient structure of the earth such as it originally was before it was changed by the motion of the waters; we should see the first state of the globe, the old matters of which it was composed, its form, and the natural arrangement of its parts; but this is too much to expect, and on too slight foundations; and it is more conformable to reason to conclude that fossil-shells are to be found in those mountains, as well as in every other place.

With respect to the manner in which shells are placed in the strata of earth or sand, Woodward says, "All shells that are met with in an infinity of strata of earth, and banks of rocks, in the highest mountains, and in the deepest quarries and mines, in flints, &c. &c. in masses of sulphur, marcasites, and other metallic and mineral bodies, are filled with similar substances to that which includes them, and never any heterogeneous matter, &c.

"In the sand stones of all countries (the specific weight of the different kinds of which vary but little, being generally with respect to water as 2-1/2 or 9/16 to 1), we find only the conchae, and other shells which are nearly of the same weight, but they are usually found in very great numbers, whereas it is very rare to meet with oyster-shells (whose specific weight is but as 2-1/3 to 1), or sea cockles (whose weight is but as 2 or 2-1/8 to 1), or other sorts of lighter shells; but on the contrary in chalk, (which is lighter than stone, being to water but as 2-1/10 to 1), we find only cockles and other kinds of lighter shells, page 32, 33."

It must be remarked, that what Woodward says in this place with respect to specific gravity, must not be looked upon as a general rule, for we find lighter and heavier shells in the same matters; for example, shells of cockles, of oysters, of echini, &c. are found in the same stones and earth; and even in the royal cabinet may be seen a petrified cockle in a cornelian, and echini petrified in an agate, &c. therefore the specific weight of the shells has not influenced so much as Woodward supposes their position in the earth. The reason why such light shells are found more abundantly in chalk is, that chalk is only the ruinated part of shells, and that those of the echini being lighter and thinner than others, would have been most easily reduced into powder or chalk, so that the strata of chalk are only met with in the places where formerly a great abundance of these light shells were collected, the destruction of which formed that chalk, in which we find those shells, which having resisted the frictions, are preserved entire, or at least in parts large enough to discover their species.

But this subject is treated more fully in our discourse on minerals; we shall here content ourselves with saying, that a modification must be given to Woodward's expressions: he seems to say, that shells are found in flints, cornelians, in ores, and sulphur, as often, and in as great a number as in other matters; whereas the truth is, that they are very rare in all vitrifiable or purely inflammable substances; and, on the contrary, are in prodigious abundance in chalk, marl, and marbles, insomuch that we cannot absolutely pretend to say, that the lightest and heaviest shells are found in corresponding strata, but only that in general they are oftener found so than otherwise. They are all filled with the substance which surrounds them, whether found in horizontal strata or in perpendicular fissures, because both have been formed by the waters, although at different times and in different manners. Those found in horizontal strata of stone, marble, &c. have been deposited by the motion of the waves of the sea, and those in flints, cornelians, and all matters which are in the perpendicular fissures, have been produced by the particular motion of a small quantity of water, loaded with lapidific or metallic substances. In both cases these matters were reduced into a fine and impalpable powder, which has filled the shells so fully and absolutely, as not to have left the least vacuum.

There is therefore in stone, marble, &c. a great multitude of shells which are whole, beautiful, and so little changed, that they may be easily compared with the shells preserved in cabinets, or found on the sea shores.

Woodward, in pages 23 and 24, proceeds, "There are, besides these, great multitudes of shells contained in stones, &c. which are entire and absolutely free from any such mineral mixture; which may be compared with those at this time seen on our shores, and which will be found not to have any difference, being precisely of the same figure and size; of the same substance and texture as the peculiar matter which composes them is the same, and is disposed and arranged in the same manner; the direction of their fibres and spiral lines are the same, the composition of the small lama formed by their fibres is the same in the one as the other; we see in the same part vestigia of tendons, by means of which the animal was fastened and joined to its shell; we see the same tubercles, stria and pipes; in short, the whole is alike, whether within or without the shell, in its cavity or on its convexity, in its substance or on its superficies. In other respects these fossil shell-fish are subject to the same common accidents as those of the sea; for example, they sometimes grow to one another, the least are adherent to the large; they have vermicular conduits; pearls are found therein, and other similar matters which have been produced by the animal when it inhabited its shell; and what is very considerable, their specific gravity is exactly the same as that of their kind found actually in the sea; in all chymical experiments they answer exactly with sea-shells; when dissolved they have the same appearance, smell and taste; in a word, their resemblance is perfectly exact."

I have often observed with astonishment, as I have already said, whole mountains, chains of rocks, enormous banks of quarries, so full of shells and other wrecks of marine productions, that their bulk surpassed that of the matter in which they were deposited.

I have seen cultivated fields so full of petrified cockles that a man might pick them up with his eyes shut, others covered with cornu ammonis, and some with cardites; and the more we examine the earth, the more we shall be convinced that the number of these petrifications is infinite, and conclude, that it is impossible that all the animals which inhabited these shells existed at one time.

I have made an observation, that in all countries where we find a very great number of petrified shells in the cultivated lands which are whole, well preserved, and totally apart, have been divided by the action of the frost, which destroys the stone and suffers the petrified shells to subsist a longer time.

This immense quantity of marine fossils found in so many places, proves that they could not have been transported thither by the deluge; for if these shells had been brought on the earth by a deluge, the greatest part would have remained on the surface of the earth, or at least would not have entered to the depth of seven or eight hundred feet in the most solid marble.

In all quarries these shells form a share of the internal part of the stone, sometimes externally covered with stalactites, which is much less ancient matter than stone, which contains shells. Another proof this happened not by a deluge is, that bones, horns, claws, &c. of land animals, are found but very rarely, and not at all in marble and other hard stone whereas if it was the effect of a deluge, where all must have perished, we should meet with the remains of land animals as well as those of the sea.

It is a vain supposition to pretend that all the earth was dissolved at the deluge, nor can we give any foundation to such idea, but by supposing a second miracle, to give the water the property of a universal dissolvent. Besides, what annihilates the supposition, and renders it even contradictory, is, that if all matters were dissolved by that water, yet shells have not been so, since we find them entire and well preserved in all the masses which are said to have been dissolved; this evidently proves that there never was such dissolution, and that the arrangement of the parallel strata was not made in an instant, but by successive sediments: for it is evident to all who will take the trouble of observing, that the arrangement of all the materials which compose the globe, is the work of the waters. The question therefore is only whether this arrangement was made at once, or in a length of time: now we have shewn it could not be done all at once, because the materials have not kept the order of specific weight, and there has not been a general dissolution; therefore this arrangement must have been produced by sediments deposited in succession of time; any other revolution, motion, or cause, would have produced a very different arrangement. Besides, particular revolutions, or accidental causes, could not have produced a similar effect on the whole globe.

Let us see what the historian of the Academy says on this subject anno 1718, p. 3. "The numerous remains of extensive inundations, and the manner in which we must conceive mountains to have been formed, sufficiently proves that great revolutions have happened to the surface of the earth. As far as we have been able to penetrate we find little else but ruins, wrecks, and vast bodies heaped up together and incorporated into one mass, without the smallest appearance of order or design. If there is some kind of regular organization in the terrestrial globe it is deeper than we have been able to examine, and all our researches must terminate in digging among the ruins of the external coat, but which will still find sufficient employment for our philosophers.

"M. de Jussieu found in the environs of St. Chaumont a great quantity of slaty or foliated stones, every foliage of which was marked with the impression of a branch, a leaf, or the fragment of a leaf of some plant: the representations of leaves were exactly extended, as if they had been carefully spread on the stone by the hand; this proves they had been brought thither by the water, which always keeps leaves in that state: they were in different situations, sometimes two or three together. It may easily be supposed that a leaf deposited by water upon soft mud, and afterwards covered with another layer of mud, imprints on the upper the image of one of its two surfaces, and on the under the image of the other; and on being hardened and petrified would appear to have taken different impressions; but, however natural this supposition may be, the fact is not so, for the two laminæ of stone bear impressions of the same side of the leaf, the one in alto, the other in bas releaf. It was M. Jussieu who made these observations on the figured stones of St. Chaumont; to him we shall leave the explication, and pass to objects which are more general and interesting.

"All the impressions on the stones of St. Chaumont are of foreign plants; they are not to be found in any part of France, but only in the East Indies or the hot climates of America; they are for the most part capillary plants, generally of the species of fern, whose hard and compact coat renders them more able to imprint and preserve themselves. Some leaves of Indian plants imprinted on the stones of Germany appeared astonishing to M. Leibnitz, but here we find the same wonderful affair infinitely multiplied. There even seems in this respect to be an unaccountable destination of nature, for in all the stones of St. Chaumont not a single plant of the country has been found.

"It is certain, by the number of fossil-shells in the quarries and mountains, that this country, as well as many others, must have formerly been covered with the sea. But how has the American or Indian sea reached thither? To explain this, and many other wonderful phenomena, it may be supposed, with much probability, that the sea originally covered the whole terrestrial globe: but this supposition will not hold good, because how were terrestrial plants to exist? It evidently, therefore, must have been great inundations which have conveyed the plants of one country into the others.

"M. de Jussieu thinks, that as the bed of the sea is continually rising, in consequence of the mud and sand which the rivers incessantly convey there, the sea, at first confined between natural dykes, surmounted them, and was dispersed over the land, and that the dykes were themselves undermined by the waters and over-thrown therein. In the earliest time of the formation of the earth, when no one thing had taken a regular form, prodigious and sudden revolutions might then have been made, of which we no longer have examples, because the whole is now in such a permanent state, that the changes must be inconsiderable and by degrees.

"By some of these great revolutions the East and West Indian seas may have been driven to Europe, and carried with them foreign plants floating on its waters, which they tore up in their road, and deposited gently in places where the water was but shallow and would soon evaporate."

ARTICLE IX.

ON THE INEQUALITIES OF THE SURFACE OF THE EARTH.

The inequalities which are on the surface of the earth, and which might be regarded as an imperfection to its figure, are necessary to preserve vegetation and life on the terrestrial globe. To be assured of this, it is only requisite to conceive what the earth would be if it was even and regular. Instead of agreeable hills, from whence pure streams of waters flow, to support the verdure of the earth; instead of those rich and flourishing meadows, where plants and animals find agreeable subsistence; a dismal sea would cover the whole globe, and the earth, deprived of all its valuable qualities, would only remain an obscure and forsaken planet, at best only destined for the abode of fishes.

But independent of moral considerations, which seldom form a proof in philosophy, there is a physical necessity why the earth must be irregular on its surface; for supposing it was perfectly regular in its origin, the motion of the waters, the subterraneous fires, the wind, and other external causes, would, in course of time, have necessarily produced irregularities similar to those now seen.

The greatest inequalities next to the elevations of mountains, are the depths of the ocean; this depth is very different even at great distances from land; it is said there are parts above a league deep, but those are few, and the most general depths are from 60 to 150 fathoms. The gulphs bordering on the coasts are much less deep, and the straits are generally the most shallow.

To sound the depths of the sea, a piece of lead of 30 or 40lb. is made use of, fastened to a small cord; this is a good method for common depths, but is not to be depended upon when the depth is considerable; because the cord being specifically lighter than the water, after it has descended to a certain degree, the weight of the lead and that of the cord is no more than a like volume of water; then the lead descends no longer, but moves in an oblique line, and floats at the same depth: to sound great depths, therefore, an iron chain is requisite, or some substance heavier than water. It is very probable that for want of considering this circumstance, navigators tell us that the sea in many places has no bottom.

In general, the profundities in open seas increase or diminish in a pretty uniform manner, and commonly the farther from shore the greater the depth; yet this is not without exception, there are places in the midst of the sea where shoals are found, as at Abrolhos in the Atlantic; and others where there are banks of a very considerable extent, as are daily experienced by the navigators to the East Indies.

So likewise along shore the depths are very unequal, nevertheless we may lay it down as a certain rule, that the depth there is always proportionate to the height of that shore. It is the same in great rivers, where the high shores always announce a great depth.

It is more easy to measure the heights of mountains, whether by means of practical geometry, or by the barometer. This instrument gives the height of a mountain very exactly, especially in a country where its variation is not considerable, as at Peru, and under the other parts of the equator. By one or other of these methods, the height of most eminences has been measured; for example, it has been found that the highest mountains of Switzerland are about 1600 fathoms higher than Canigau, which is one of the most elevated of the Pyrennees; those mountains appear to be the highest in Europe, since a great quantity of rivers flow from them, which carry their water into very remote and different seas, as the Po, which flows into the Adriatic; the Rhine, which loses itself in the sands in Holland; the Rhone, which falls into the Mediterranean; and the Danube, which goes to the Black Sea. These four rivers, whose mouths are so remote from each other, all derive a part of their waters from Mount Saint Godard and the neighbouring mountain, which proves that this place is the highest in all Europe. The highest mountains in Asia are Mount Taurus, Mount Imaus, Caucasus, and the mountains of Japan, all which are loftier than those of Europe; the mountains in Africa, as the Great Atlas, and the mountains of the Moon, are at least as high as those in Asia, and the highest of all are in South America, particularly those of Peru, which are more than 3000 fathoms above the level of the sea. In general the mountains between the tropics are loftier than those of the temperate zones, and these more than the frigid zones, so that the nearer we approach the equator, the greater are the inequalities of the earth. These inequalities, although very considerable with respect to us, are scarcely any thing when considered with respect to the whole globe. Three thousand fathom difference to 3000 leagues diameter, is but one fathom to a league, or one foot to 2200 feet, which on a globe of 2-1/2 feet diameter, does not make the 16th part of a French line. Thus the earth, which appears to us crossed and intersected by the enormous height of mountains, and by a frightful depth of sea, is nevertheless, relative to its size, but slightly furrowed with irregularities, so very trifling, that they can cause no difference to the general figure of the globe. In continents the mountains are continued and form chains. In islands, they are more interrupted, and generally raised above the sea, in the forms of cones or pyramids, and are called peaks. The peak of Teneriffe, in the island of Fer, is one of the highest mountains on the earth; it is near a league and a half perpendicular above the level of the sea; the peak of St. George, in one of the Azores, and the peak of Adam, in the island of Ceylon, are also very lofty. These peaks are composed of rocks, heaped one upon the other, and they vomit from their summits fire, cinders, bitumen, minerals, and stones. There are islands which are only tops of mountains, as of St. Helena, Ascension, most of the Azores, and Canaries. We must remark, that in most of the islands, promontories, and other projecting lands in the sea, the middle is always the highest; and they are generally separated by chains of mountains, which divide them in their greatest length, as (Gransbain) the Grampian mountains in Scotland, which extend from east to west, and divide

Great Britain into two parts. It is the same with the islands of Sumatra, Lucca, Borneo, Celebes, Cuba, St. Domingo, and the peninsula of Malaya, &c. and also Italy, which is traversed through its whole length by the Apennine mountains.

Mountains, as we find, differ greatly in height; the hills are lowest, after them come the mountains of a moderate height, which are followed by a third rank still higher, which, like the preceding, are generally loaded with trees and plants, but which furnish no springs except at their bottoms. In the highest mountains we find only sand, stones, flints, and rocks, whose summits often rise above the clouds. Exactly at the foot of these rocks there are small spaces, plains, hollows, and kinds of vallies, where the rain, snow, and ice remain, and form ponds, morasses, and springs, from whence rivers derive their origin.

The form of mountains is also very different: some form chains whose height is nearly equal in a long extent of soil, others are divided by deep vallies; some are regular, and others as irregular as possible; and sometimes in the middle of a valley or plain, we find a little mountain. There are also two sorts of plains, the one in the low lands, the other in mountains. The first are generally divided by some large river: the others, though of a very considerable extent, are dry, and at farthest have only a small rivulet. These plains on mountains are often very high, and difficult of access; they form countries above other countries, as in Auvergne, Savoy, and many other high places: the soil is firm, and produces much grass, and odoriferous plants, which render these plains the best pasture in the world.

The summits of high mountains are composed of rocks of different heights, which resemble from a distance the waves of the sea. It is not on this observation alone we can rely that the mountains have been formed by the waves, I only relate it because it accords with the rest: but that which evidently proves that the sea once covered and formed mountains, are the shells and other marine productions found throughout in such great quantities, that it is not possible for them to have been transported by the sea into such remote continents, and deposited to such considerable depths; to this may be added, the horizontal and parallel strata every where met with, and which can only have been formed by the waters. The composition even of the hardest matters, as stone and marble, prove they had been reduced into fine powder before their formation, and precipitated to the bottom of the water in form of a sediment: it is also proved by the exactness with which fossil-shells are moulded in those matters in which they are found; the inside of these shells are absolutely filled with the same matters as that in which they are enclosed; the corresponding angles of mountains and hills, which no other cause than the currents of the sea could have been able to form; the equality in the height of opposite hills, and beds of different matters, formed at the same levels, and, in short, the direction of mountains, whose chains extend in length in the same direction as the waves of the sea extend, incontestibly demonstrate the fact.

With respect to the depths on the surface of the earth, the greatest, without contradiction, are the depths of the sea; but as they do not present themselves to our sight, and as we can only judge of them by the plumb line, we shall only speak of those which appear on dry land, such as the deep vallies between mountains, the precipices between rocks, the abysses perceived from the tops of mountains, as the abyss of Mount Ararat, the precipices of the Alps, the vallies of the Pyrennees, &c. These depths are a natural consequence of the elevation of mountains; they receive the waters and the earth which flow from the mountains, and the soil is generally very fertile, and are fully inhabited.

The precipices which are between rocks are frequently formed by the sinking of one side, the base of which sometimes gives way more on one side than the other, by the action of the air and frost, which splits and divides them, or by the impetuous violence of torrents. But these abysses, or vast and enormous precipices, found at the summits of mountains, and to the bottom of which it is not possible sometimes to descend, although they are above a mile, or a mile and a half round, have been formed by the fire. These were formerly the funnels of volcanos, and all the matter which is there deficient has been ejected by the action and explosion of these fires, which are since extinguished through a defect of combustible matter. The abyss of Mount Ararat, of which M. Tournefort gives a description in his voyage to the Levant, is surrounded with black and burnt rocks, as one day the abysses of Etna, Vesuvius, and other volcanos, will be, when they have consumed all the combustible matters they include.

In Plots' Natural History of Staffordshire, in England, a kind of gulph is spoken of which has been sounded to the depth of 2600 perpendicular feet without meeting with any water, or the bottom being found, as the rope was not of sufficient length to reach it.

Greatest cavities and deepest mines are generally in mountains, and they never descend to a level with the plains, therefore by these cavities we are only acquainted with the inside of a mountain, and not with the internal part of the globe itself.

Besides, these depths are not very considerable. Ray asserts that the deepest mines are not above half a mile deep. The mine of Cotteberg, which in the time of Agricola passed for the deepest of all known mines, was only 2500 feet perpendicular. It is evident there are holes in certain places, as that in Staffordshire, or Pool's Hole, in Derbyshire, the depth of which is perhaps greater; but all this is nothing in comparison with the thickness of the globe.

If the kings of Egypt, instead of having erected pyramids, and raised such sumptuous monuments of their riches and vanity, had been at the same expence to sound the earth, and make a deep excavation to the depth of a league, they, perhaps, might have found substances which would have amply recompensed the trouble, labour, and expence, or at least we should have received information on the matters of which the internal part of the globe is composed, which might have been very useful, and which we at present have not.

But let us return to the mountains; the highest are in the southern countries, and the nearer we approach the equator, the more inequalities we find on the surface of the globe. This is easy to prove, by a short enumeration of the mountains and islands.

In America, the chain of the Cordeliers, the highest mountains of the earth, is exactly under the equator, and extends on the two sides far beyond the tropic circles.

In Africa, the highest mountains of the Moon, and Monomotapa, the great and the little Atlas, are under the equator, or not far from it.

In Asia, Mount Caucasus, the chain of which extends under different names as far as the mountains of China, is nearer the equator than the poles.

In Europe, the Pyrennees, the Alps, and mountains of Greece, which are only the same chain, are still less distant from the equator than the poles.

Now these mountains which we have enumerated, are all higher, more considerable and extended in length and breadth than the mountains of the northern countries.

With respect to their direction, the Alps form a chain which crosses the whole continent from Spain to China. These mountains begin at the sea coast of Galicia, reach to the Pyrennees, cross France, by Vivares, and Auvergne, pass through Italy and extend into Germany, beyond Dalmatia, as far as Macedonia; from thence they join with the mountains of Armenia, Caucasus, Taurus, Imaus, and extend as far as the Tartarian sea. So likewise Mount Atlas traverses the whole continent of Africa, from west to east, from the kingdom of Fez to the Straits of the Red Sea; and the mountains of the Moon have the same direction.

But in America, the direction is quite contrary, and the chains of the Cordeliers and other mountains extend from south to north more than from east to west.

What we have now said on the great eminences of the earth, may also be observed on the greatest depths of the sea. The vast and highest seas are nearer the equator than the poles; and there results from this observation, that the greatest inequalities of the globe are in the southern climate. These irregularities on the surface of the earth, are the causes of an infinity of extraordinary effects: for example, between the Indus and the Ganges, there is a large peninsula, which is divided through its middle, by a chain of high mountains called the Gate, and which extends from north to south, from the extremities of Mount Caucasus to Cape Comorin; on one is the coast of Malabar, and the other Coromandel; on the side of Malabar, between this chain of mountains and the sea, the summer season lasts from September to April, during which the sky is serene and dry; on the other side the Coromandel the above period is their winter, and it rains every day plentifully and from the month of April to the month of September is their summer, whereas it is winter in Malabar; insomuch, that in many places, which are scarcely 20 miles distant, we may, by crossing the mountains, change seasons. It is said that the same thing takes place at Razalgat in Arabia, and at Jamaica, which is divided through its middle by a chain of mountains, whose direction is from east to west, and that the plantations to the south of these mountains feel the summer heat, at the time those to the north endure the rigor of winter.

Peru, which is situated under the line, and extends about a thousand leagues to the south, is divided into three long and narrow parts; these the natives call Lanos, Sierras, and Andes. The Lanos, which comprehends the plains, extends along the coast of the South Sea: the Sierras are hills with some vallies, and the Andes are the famous Cordeliers, the highest mountains that are known. The Lanos is about ten leagues in breadth; in many places the Sierras are twenty leagues broad, and the Andes in some places more and in some less. The breadth is from east to west, and the length from north to south. This part of the world is remarkable for the following particulars: first, in the Lanos the wind almost constantly blows from the south-west, which is contrary to what happens in the torrid zone: secondly, it never rains nor thunders in the Lanos, although there is plenty of dew: thirdly, it almost continually rains in the Andes: fourthly, in the Sierras, between the Lanos and the Andes, it rains from September to April.

It was for a long time supposed, that the chains of the high mountains run from west to east, till the contrary was found in America. But no person before M. Bourguet discovered the surprising regularity of the structure of those great masses: he found (after having crossed the Alps thirty times in fourteen different parts of it, twice over the Apennine mountains, and made divers tours in the environs of these mountains, and of Mount Jura) that all mountains are formed nearly after the manner of works of a fortification. When the body of the mountain runs from east to west, it forms prominences, which face the north and south; this wonderful regularity is so striking in vallies, that we seem to walk in a very regular covered way; if, for example, we travel in a valley from north to south, we perceive that the mountain on the right forms projections which front the east, and those of the mountain on the left front the west, so that the saliant angles of one side reciprocally answer the returning angles of the other, which are always alternatively opposed to them. The angles which mountains form in great vallies are less acute, because the direction is less steep, and they are farther distant from each other. In plains they are not so perceptible, except by the banks of rivers, which are generally in the middle of them, and whose natural windings answer the most advanced angles or striking projections of the mountains. It is astonishing so visible a thing was so long unobserved, for when in a valley the inclination of one of the mountains which border it is less steep than that of the other, the river takes its course much nearer the steepest mountain, and does not flow through its middle.

To these observations we may join other particular ones, which confirm them; for example, the mountains of Switzerland are much more steep, and their direction much greater on the south side than on the north, and on the west side than on the east. This may be perceived in the mountains of Gemmi, Brisa, and almost every other mountain in this country. The highest are those which separate Valesia and the Grisons from Savoy, Piedmont, and Tirol. These countries are only a continuation of these mountains, the chain of which extends to the Mediterranean, and continues even pretty far under the sea. The Pyrennees are also only a continuation of that vast mountain which begins in Upper Valesia, and whose branches extend very far to the west and south, preserving throughout the same great height; whereas on the side of the north and of the east these mountains grow lower by degrees, till they become plains; as we see by the large tract which the Rhine and Danube water before they reach their mouths, whereas the Rhone descends with rapidity towards the south into the Mediterranean. The same observation is found to hold good in the mountains of England and Norway; but the part of the world where this is most evidently seen is at Peru and Chili; the Cordeliers are cut very sharply on the western side, the length of the Pacific Ocean, whereas on the eastern side they lower by degrees into large plains, watered by the greatest rivers of the world.[32]

M. Bourguet, to whom we owe this great discovery of the correspondence of the angles of mountains, terms it *"The Key of the Theory of the Earth;"* nevertheless, it appears to me, that if he had conceived all the importance of it, he would more successfully have made use of it, by connecting it with suitable facts, and would have given a more probable theory of the earth; whereas in his treatise he presents only the skeleton of an hypothetical system, most of the conclusions of which are false or precarious. The theory we have given turns on four principal facts, which cannot be doubted, after the proofs have been examined on which they are founded. The first is, that the earth is every where, and to considerable depths, composed of parallel strata, and matters which have formerly been in a state of softness: the second, that the sea has for ages covered the earth which we now inhabit; the third, that the tides and other motions of the waters produce inequalities at the bottom of the sea; and the fourth, that the mountains have taken their form and the correspondent direction from the currents of the sea.

After having read the proofs which the following articles contain, it may be determined, whether I was wrong to assert, that these circumstances solidly established also ascertains the truth of the theory of the earth. What I have said on the formation of mountains has no need of a more ample explanation; but as it might be objected that I do not assign a reason for the formation of the peaks or points of mountains, no more than for some other particular circumstances, shall add the observations and reflections which I have made on this subject.

I have endeavoured to form a clear and general idea of the manner in which the different matters that compose the earth are arranged, and it appears to me they may be reduced into two general classes; the first includes all the matters we find placed in strata, or beds horizontally or regularly inclined; and the second comprehends all matters formed in masses, or in veins, either perpendicular or irregularly inclined. In the first class are included sands, clays, granite, flints, free-stone, coals, slates, marls, chalks, calcinable stones, marbles, &c. In the second I rank metals, minerals, crystals, precious stones and small flints: these two classes generally comprehend all the known materials of the earth. The first owe their origin to the sediments carried away and deposited by the sea, and should be distinguished into those which being assayed in the fire, calcine and are reduced into lime, and those which fuse and are convertible into glass. The materials of the second class are all vitrifiable excepting those which the fire entirely consumes by inflammation.

In the first class we distinguish two kinds of sands; the one, which is more abundant than any other matter of the globe, is vitrifiable, or rather is only fragments of actual glass; the other, whose quantity is much less, is calcinable, and must be looked upon as the powder of stone, and which differs only from gravel by the size of the grains. The vitrifiable sand is, in general, deposited in beds,

[32] See Phil. Trans. Abr. Vol. VI. part ii. p. 153.

which are often interrupted by masses of free-stone, granite, and flint; and sometimes these matters are also in banks of great extent.

By examining these vitrifiable matters, we find only a few sea shells there, and those not placed in beds, but dispersed about as if thrown there by chance. For example, I have never seen them in free-stone; that stone which is very plenty in certain places, is only composed of sandy parts, which are re-united, and are only met with in sandy soils; and the quarries of it are generally in peaked hills and in divided eminences. We may work these quarries in all directions, and if they are in large beds, they are much farther from each other than in quarries of calcinable stone or marble. Blocks of free-stone may be cut of all dimensions and in all directions, although it is difficult to work, it nevertheless has but a degree of hardness sufficient to resist powerful strokes without splitting; for friction easily reduces it into sand, excepting certain black pieces found therein, and which are so very hard, that the best files cannot touch them. Rock is vitrifiable as free-stone, and of the same nature, only it is harder and the parts more connected. This also contains many hard pieces, as may easily be remarked on the summits of high mountains, which cut and tear the shoes of travellers. This rocky stone, which is found at the top of high mountains, and which I look upon as a kind of granite, contains a great quantity of talky leaves, and is so hard as not to be worked but by an infinite deal of labour.

I have narrowly examined these sharp pieces which are found in free-stone and rock, and have discovered it to be a metallic matter, melted and calcined by a very violent fire, and which perfectly resembles certain substances thrown out by the volcanos, of which I saw a great quantity when I was in Italy, where the people called them Schiarri. They are very heavy black masses, on which neither water nor the file can make any impression, and the matter of which is different from that of the lava; for this is a kind of glass, whereas the other appears to be more metallic than vitreous. The sharp pieces in free-stone, and rock, resemble greatly the first matter, which seems still to prove that all these matters have been formerly liquified by fire.

We sometimes see on the upper parts of mountains, a prodigious quantity of blocks of this mixed rock; their position is so irregular that they appear to have been thrown there by chance, and it might be thought they had fallen from some neighbouring height, if the places where they are found were not raised above the other parts. But their vitrifiable nature, and their angular and square figures, like those of free-stone, discover them to be of one common origin. Thus in the great beds of vitrifiable sand, blocks of free-stone and rock are formed, whose figures and situations do not exactly follow the horizontal position of these strata. The rain, by degrees, carried away from the summits of the hills and mountains the sand which at first covered them, and then began to furrow and cut those hills into the spaces which are found between the nucleus in free-stone, as the hills of Fontainbleau are intersected. Each hilly point answers to a nucleus in a quarry of free-stone, and each interval has been excavated and loosened by the rain, which has caused the sand, they at first contained, to flow into the vallies; so likewise the highest mountains, whose summits are composed of rocks, and terminated by these angular blocks of granite, have formerly been covered with vitrifiable sand, and the rain having carried away the sand which covered them, they remained on the tops of the mountains in the position they were formed. These blocks generally present points; they increase in size in proportion as they descend; one block often rests upon another, the second upon a third, and so on, leaving irregular intervals between them: and as in time the rain washed away all the sand which covered these different parts on the top of the high mountains, they would remain naked, forming larger or lesser points; and this is the origin of the peaks or horns of mountains.

For supposing, as it is easy to prove by the marine productions we find there, that the chain of the Alps was formerly covered by the sea, and that above this chain there was a great thickness of vitrifiable sand, which rendered the whole mountains a flat and level country. In this depth of sand, there would necessarily be formed granite, free-stone, flint, and all matters which take their origin and figure in sand, nearly in a similar manner to that of the crystallisation of salts. These blocks once formed would support their original positions, after the rains and torrents had carried away the sand which surrounded them, and being left bare formed all those peaks or pointed eminences we see in so many places. This is also the origin of those high and detached rocks found in China and other countries, as in Ireland, where they are called the Devil's stones, and whose formation as well as that of the peaks of mountains, had hitherto appeared so difficult to explain; nevertheless the explanation which I have given is so natural, that it directly presents itself to the mind of those who examine these objects, and I must here quote what Father Tatre says, "From Yanchu-in-yen, we came to Hoytcheou, and on the road met with something particular, rocks of an extraordinary height, of the shape of a large square tower, and situate in the midst of vast plains: I cannot account for it, unless by supposing they were formerly mountains, from which the rain having washed away the earth that surrounded them, thus left the rocks entirely bare. What strengthens this conjecture is, that we saw some which, towards the base, are still covered with earth to a considerable height."

The summits of the highest mountains are composed of rocks, of granite, free-stone, and other hard and vitrifiable matters, and this often as deep as two or three hundred fathoms; below which we often meet with quarries of marble, or hard stone, filled with fossil-shells, and whose matter is calcinable; as may be remarked at Great Chartreuse, in Dauphiny, and on Mount Cenis, where the stone and marble, which contains shells, are some hundred fathoms below the summits, points and peaks of high mountains; although these stones are more than a thousand fathom above the level of the sea. Thus mountains, whereon we see points or peaks, are generally vitrifiable rock, and those whose summits are flat, mostly contain marble and hard stones filled with marine productions. It is the same with respect to hills, for those containing granite, or free-stone, are mostly intersected with points, eminences, cavities, depths, and small intermediate valleys; on the contrary, those which are composed of calcinable stone are nearly equal in height, and are only interrupted by greater and more regular vallies, whose angles are correspondent; and they are crowned with rocks whose position is regular and level.

Whatever difference may appear at first between these two species of mountains, their forms proceed from the same cause, as we have already observed; only it may be remarked, that the calcinable stones have not undergone any alteration nor change since the formation of the horizontal strata; whereas those of vitrifiable sand have been changed and interrupted by the posterior production of rocks and angular blocks formed within this sand. These two kinds of mountains have cracks which are almost always perpendicular in those of calcinable stones; but those of granite and free-stone appear to be a little more irregular in their direction. It is in these cracks metal, minerals, crystals, sulphurs, and all matters of the second class are found, and it is below these cracks that the water collects to penetrate the earth, and form those veins of water which are every where found below the surface.

ARTICLE X.

OF RIVERS.

We have before said that, generally speaking, the greatest mountains are in islands and in the projections in the sea. That in the old continent the greatest chains of mountains are directed from west to east, and that those which incline towards the north or south are only branches of these principal chains; we shall likewise find that the greatest rivers are directed as the greatest mountains, and that there are but few which follow the course of the branches of those mountains. To be assured of this, we have only to look on a common globe, and trace the old continent from Spain to China. We shall find, by beginning at Spain, that the Vigo, Douro, Tagos, and Guadiana run from east to west, and the Ebro from west to east, and that there is not one remarkable river whose course is directed from south to north, or from north to south, although Spain is entirely surrounded by the sea on the west side, and almost so on the north. This observation on the directions of rivers in Spain not only proves that the mountains in this country are directed from west to east, but also that the southern lands, which border on the straits, are higher than the coasts of Portugal; and on the northern coast, that the mountains of Galicia, the Asturias, &c. are only a continuation of the Pyrennees, and that it is this elevation of the country, as well north as south, which does not permit the rivers to run into the sea that way.

It will also be seen, by looking on the map of France, that there is only the Rhone which runs from north to south, and nearly half its course, from the mountains to Lyons, is directed from the east towards the west; but that on the contrary all the other great rivers, as the Loir, the Charantee, the Garonne, and even the Seine, have a direction from east to west.

It will be likewise perceived, that in Germany there is only the Rhine, which like the Rhone shapes the greatest part of its course from north to south, but that the others, as the Danube, the Drave, and all the great rivers which fall into them, flow from the west to east into the Black Sea.

It will be perceived that this Black Sea, which should rather be considered as a great lake, has almost three times more extent from east to west than from north to south, and consequently its direction is similar to the rivers in general. It is the same with the Mediterranean, whose length from east to west is about six times greater than from north to south.

The Caspian Sea, according to the chart drawn by the order of Czar Peter I. has more extent from the south to the north than from east to west; whereas in the ancient charts it appears almost round, or rather more broad from east to west than from south to north; but if we consider the lake Aral as a part of the Caspian Sea, from which it is separated only by plains of sand, we shall find the length is from the western coast of the Caspian Sea as far as the greatest border of Lake Aral.

So likewise the Euphrates, the Persian gulph, and almost all the rivers in China run from west to east; all the rivers in Africa beyond Barbary flow from east to west, or from west to east, and there are only the rivers of Barbary and the Nile which flow from south to north. There are, in fact, great rivers in Asia which partly run from north to south, as the Wolga, the Don, &c. but by taking the whole length of their course, we find, that they only turn from the south to run into the Black and Caspian seas, which are only inland lakes.

It may therefore in general be said, that in Europe, Asia, and Africa, the rivers, and other mediterranean waters, extend more from east to west than from north to south, which proceeds from the chains of mountains being for the most part so directed, and that the whole continent of Europe and Asia is broader in this direction than the other; for there are two modes of considering the direction of mountains. In a long and narrow continent like South America, in which there is only one principal chain of mountains which stretches from south to north, the river not being confined by any parallel range, necessarily runs perpendicular to the course of the mountains, that is from east to west, or from west to east; in fact, it is in this direction all the rivers of America flow. In the old as well as the new continent most of the waters have their greatest extent from west to east, and most of the rivers flow in this direction; but yet this similar direction is produced by different causes; for instance, those in the old continent flow from east to west, because they are bounded by mountains whose direction is from west to east; whereas those in America preserve the same course from there being only one chain of mountains that extends from north to south.

In general, rivers run through the centre of vallies, or rather the lowest ground betwixt two opposite hills or mountains; if the two hills have nearly an equal inclination, the river will be nearly in the middle of the intermediate valley, let the valley be broad or narrow. On the contrary, if one of the hills has a more steep inclination than the other, the river will not be in the middle of the valley, but much nearer the hill whose inclination is greatest, and that too in proportion to the superiority of its declivity: in this case, the lowest ground is not in the middle of the valley, but inclines towards the highest hill, and which the river must necessarily occupy. In all places where there is any considerable difference in the height of the mountains, the rivers flow at the foot of the steepest hills, and follow them throughout all their directions, never quitting their course while they maintain the superiority of height. In the length of time, however, the steepest hills are diminished by the rain acting upon them with a greater degree of force, proportionate to their height, and consequently carry away the sand and gravel in more considerable quantities, and with greater violence; the river is then constrained to change its bed, and seek the lowest part of the valley: to this may be added, that as all rivers overflow at times, they transport and deposit mud and sand in different places, and that sands often accumulate in their own beds, and cause a swell of the water, which changes the direction of its course. It is very common to meet in vallies with a great number of old channels of the river, particularly if it is subject to frequent inundations, and carries off much sand and mud.

In plains and large vallies, where there are great rivers, the beds are generally the lowest part of the valley, but the surface of the water is very often higher than the ground adjacent. For example, when a river begins to overflow, the plain will presently be inundated to a considerable breadth, and it will be observed that the borders of the river will be covered the last; which proves that they are higher than the rest of the ground, and that from the banks to a certain part of the plain, there is an insensible inclination, so that the surface of the water must be higher than the plain when the river is full. This elevation on the banks of rivers proceeds from the deposit of the mud and sand at the time of inundations. The water is commonly very muddy in the great swellings of rivers; when it begins to overflow, it runs very gently over the banks, and by depositing the mud and sand purifies itself as it advances into the plain; so that all the soil which the currents of the river does not carry along, is deposited on the banks, which raises them by degrees above the rest of the plain.

Rivers are always broadest at their mouths; in proportion as we advance in the country, and are more remote from the sea, their breadth diminishes; but what is more remarkable, in the inland parts they flow in a direct line, and in proportion as they approach their mouths the windings of their course increase. I have been informed by M. Fabry, a sensible traveller, who went several times by land into the western part of North America, that travellers, and even the savages, are seldom deceived in the distance they are from the sea if they follow the bank of a large river; when the direction of the river is straight for 15 or 20 leagues, they know themselves to be a great distance from the coast; but, on the contrary, if the river winds, and often changes its direction, they are certain of not being far from the sea. M. Fabry himself verified this remark in his travels over that unknown and almost uninhabited country. In large rivers there is a considerable eddy along the banks, which is so much the more considerable as the river is less remote from the sea, which may also serve as a guide to judge whether we are at a great or short distance from the mouth; and as the windings of rivers increase in proportion as they approach the sea, it is not surprising that some of them should give way to the water, and be one reason why great rivers generally divide into many arms before they gain the sea.

The motion of the waters in rivers is quite different from that supposed by authors who attempt to give mathematical theories on this subject; the surface of a river in motion is not level when taken from one bank to the other, but according to circumstances the current in the middle is considerably higher or lower than the water close to the banks; when a river swells by a sudden melting of snow, or when by some other cause its rapidity is augmented, if the direction of the river is straight, the middle of the water where the current is rises, and the river forms a convex curve, of a very sensible elevation. This elevation is sometimes very considerable; M. Hupeau, an able engineer of bridges, once measured the river Avieron, and found the middle was three feet higher than near the bank. This, in fact, must happen every time the water has a very great rapidity; the velocity with which it is carried, diminishing the action of its weight in the middle of the current, so that it has not time to sink to a level with that near shore, and therefore remains higher. On the other hand, near the mouths, it often happens that the water which is near the banks is higher than that of the middle, although the current be ever so rapid. This happens wherever the action of the tides is felt in a river, which in great ones often sensibly extends as far as one or two hundred leagues from the sea; it is also a well known fact that the current of a river preserves its motion in the sea to a considerable distance; there is, in this case, therefore, two contrary motions in a river; the middle, which forms the current, precipitates itself towards the sea, and the action of the tide forms a counter-current, which causes the water near the banks to ascend, while that in the middle descends, and as then all the water must be carried down by the current in the middle, that of the banks continually descends thereto, and descends so much the more as it is higher, and counteracted with more force by the tide.

There are two kinds of ebbings in rivers; the first above-mentioned is a strong power occasioned by the tide, which not only opposes the natural motion of the river, but even forces a contrary and opposite current. The other arises from an inactive cause, such as a projection of land, an island, &c. This does not commonly occasion a very sensible counter-current, yet it is sufficient to impede the progress of boats and craft, and necessarily produces what is called a dead water, which does not flow like the rest of the river, but whirls about in such a manner that when boats are drawn therein they require great strength to get them out. These dead waters are very perceptible at the arches of bridges in rapid rivers. The velocity of the water increases in proportion as the diameter of the channel through which it passes diminishes, the impelling force being the same; the velocity of a river, therefore, increases at the passage of a bridge, in an inverse proportion of the breadth of the arches to the whole breadth of the river; the rapidity being very considerable in coming through the arch, it forces the water against the banks, from whence it is reflected with such violence as to form dangerous eddies and whirlpools. In going through the bridge St. Esprit, the men are forced to be careful not to lose the stream, even after they are past the bridge, for if they suffer the boat to go either to the right or left, it might be driven against the shore, or forced into the whirling waters, which would be attended with great danger. When this eddy is very considerable, it forms a kind of small gulph, the middle of which appears hollow and to form a kind of cylindrical cavity, around which the water whirls with rapidity: this appearance of a cylindrical cavity is produced by the centrifugal force, which causes the water to endeavour to remove itself from the centre of the whirlpool. When a great swell of water happens, the watermen know it by a particular motion; they then say the water at the bottom flows quicker than common: this augmentation of rapidity at the bottom, according to them, always announces a sudden rise of the water. The motion and weight of the upper water communicates this motion to them; for in certain respects we must consider a river as a pillar of water contained in a tube, and the whole channel as a very long canal where every motion must be communicated from one end to the other. Now, independent of the motion of the upper waters, their weight alone might cause the rapidity of the river to increase, and perhaps move it at bottom; for it is known, that by putting many boats at one time into the water, at that instant we increase the rapidity of the under part of the river, as well as retard that of the upper.

The rapidity of running waters does not exactly, nor even nearly, follow the proportion of the declivity of their channels. One river whose inclination is uniform and double that of another, ought, according to appearance, to flow only as rapid again, but in fact it flows much faster. Its rapidity, instead of being doubled, is sometimes triple, quadruple, &c. This rapidity depends much more on the quantity of water and the weight of the upper waters than on the declivity. When we are desirous to hollow the bed of a river, we need not equally distribute the inclination throughout its whole length, in order to give a greater rapidity, as it is more easily effected by making the descent much greater at the beginning, than at the mouth, where it may almost be insensible, as we see it in natural rivers, and yet they preserve a rapidity so much the greater as the river is fuller of water; in great rivers, where the ground is level, the water does not cease flowing, and even rapidly, not only with its original velocity, but also with the addition of that which it has acquired by the action and weight of the upper waters. To render this fact more conceivable, let us suppose the Seine between the Pont-neuf and Pont-royal to be perfectly level, and ten feet deep throughout: let us then suppose that the bed of the river below Pont-royal and above Pont-neuf were left entirely dry, the water would instantly run up and down the channel, and continue to do so until it had recovered an equilibrium; for the weight of the water would keep it in motion, nor would it cease flowing until its particles became equally pressed and have sunk to a perfect level. The weight of water therefore greatly contributes to its velocity, and this is the reason that the greatest rapidity of the current is neither of the surface nor at the bottom of the water, but nearly in the middle of its depth, being pressed by the action of its weight at its surface, and by the re-action from the bottom. Still more, if a river has acquired a great rapidity, it will not only preserve it in passing a level country, but even surmount an eminence without spreading much on either side, or at least without causing any great inundation.

We might be inclined to think that bridges, locks, and other obstacles raised on rivers, considerably diminishes the celerity of the water's course; nevertheless that occasions but little difference. Water rises on meeting with any obstacle, and having surmounted it, the elevation causes it to act with more rapidity in its fall, so that in fact it suffers little or no diminution in its celerity, by these seeming retardments. Sinuosities, projections, and islands, also but very little diminish the velocity of the course of rivers. A considerable diminution is produced by the sinking of the water, and, on the contrary, its augmentation increases its velocity; thus if a river is shallow the stream passes slowly along, and if deep with a proportionate rapidity.

If rivers were always nearly of an equal fulness, the best means of diminishing their rapidity, and confining them within their banks, would be to enlarge their channel; but as almost all rivers are subject to increase and diminish, to confine them we must retrench the channel, because in shallow waters, if the channel is very broad, the water which passes in the middle hollows a winding bed, and when it begins to swell follows the direction it took in this particular bed, and striking forcibly against the banks of the channel destroys them and does great injuries. These effects of the water's fury might be prevented by making, at particular distances, small gulphs in the earth; that is, by cutting through one of these banks to a certain distance in the land. In order that these gulphs might be advantageously placed, they should be made in the obtuse angle of the river, for then the current of the water in turning would run into them, and of course its velocity would be diminished. This mode might be proper to prevent the fall of bridges in places where it is not possible to make bars near the bridge which sustain the action of the weight of the water.

The manner in which inundations are occasioned merits peculiar attention. When a river swells, the rapidity of the water always increases till it begins to overflow the banks; at that instant the velocity diminishes, which causes inundations to continue for several days; for when even a less quantity of water comes after the overflowing than before, the inundation will still be made, because it depends much more on the velocity of the water than on the quantity; if it was not so rivers would overflow for an hour or two and then return to their beds, which never occurs; the inundations always remaining for several days; whether the rain ceases, or a less quantity of water is brought, because the overflowing has diminished the velocity, and consequently, although the like quantity of water is no longer carried in the same time as before, yet the effects are the same as if the greater quantity had come there. It might be remarked on the occasion of this diminution, that if a constant wind blows against the current of the river, the inundation will be much greater than it would have been without this accidental cause, which diminishes the celerity of the water; on the contrary, if the wind blows in the same directions with the current, the inundation will be much less, and will more speedily decline.

"The swelling of the Nile, says M. Granger, and its inundations, has a long time employed the learned; most of them have looked upon it as marvellous, although nothing can be more natural, and is every day to be seen in every country throughout the world. It is the rains which fall in Abyssinia and Ethiopia which cause the swelling and inundation of that river, though the north wind must be regarded as the principal cause. 1. Because the north wind drives the clouds which contain this rain into Abyssinia. 2. Because, blowing against the mouths of the Nile, it causes the waters to return against the stream, and thus prevents them from running out in any great quantity: this circumstance may be every season observed, for when the wind, being at the north, suddenly veers to the south, the Nile loses in one day more than it gathers in four."

Inundations are generally greatest in the upper part of rivers, because the velocity of a river continues always increasing until it arrives at the sea, for the reasons we have related. Father Costelli, who has written very sensibly on this subject, remarks, that the height of the banks made to confine the Po from overflowing diminishes as they advance towards the sea; so that at Ferrara, which is 50 or 60 miles from the sea, they are near 20 feet high above the common surface of the Po, but that at 10 or 12 miles from it they are not above 12 feet, although the channel of the river is as narrow there as at Ferrara[33].

On the whole, the theory of the motion of running waters is still subject to many difficulties, nor is it easy to lay down rules which might be applied to every particular case. Experience is here more useful than speculation. We must not only know the general effects of rivers, but we must also know in particular the river we have to do with, if we would reason justly, make useful observations, and draw stable conclusions. The remarks I have above given are mostly new; it is to be wished that others may be collected, and then, possibly, in time, we may obtain a sufficient knowledge of the subject to lay down certain rules to confine and direct rivers, and prevent the ruin of bridges, banks, and other damages which the violent impetuosity of the water occasions.

The greatest rivers in Europe are the Wolga, which is about 650 leagues in its course from Reschow to Astracan, on the Caspian Sea; the Danube, whose course is about 450 leagues from the mountains of Switzerland to the Black Sea; the Don, which is 400 leagues in its course from the source of the Sosnia, which it receives, to its mouth in the Black Sea; the Dnieper, whose course is about 350 leagues, and which also runs into the Black Sea; the Duine is about 400 leagues in its course, and empties itself into the White Sea, &c.

The greatest rivers in Asia are the Hoanho of China, whose course is 850 leagues, taking its source at Raja-Ribron, and falls into the sea of China, in the middle of the gulph Changi: the Jenisca of Tartary, which is about 800 leagues in extent, from the lake Seligna to the northern sea of Tartary; the river Oby, which is about 600 leagues from Lake Kila, to the Northern Sea, beyond the Strait of Waigats. The river Amour, of eastern Tartary, which is about 575 leagues in its course, reckoning it from the source of the river Kerlon, to the sea of Kamschatka. The river Menan, whose mouth is at Poulo Condor, may be measured from the surface of the Longmu which falls into it; the Kian, whose course is about 550 leagues from the source of the river Kinxa, which it receives, to its mouth in the China Sea; the Ganges is also about 550 leagues, and the Euphrates 500, taking it from the source of Irma, which it receives. The Indus about 400 leagues, and which falls into the Arabian Sea, on the east of Guzarat. The Sirderious, which is about 400 leagues long, and falls into Lake Aral.

The greatest rivers in Africa are Senegal, which is 1125 leagues long, comprehending the Niger, which in fact is a continuation of it, and the source of Gombarou, which falls into the Niger. The Nile 970 leagues long, and which derives its source in Upper Ethiopia, where it makes many windings. There are also the Zaira, the Coanza, and the Couma, which are known as far as 400 leagues, but extend much farther; the Quilmanci, whose course is 400 leagues, and which derives its source in the kingdom of Gingiro.

The greatest rivers of America, and which are also the greatest in the world, are the river Amazons, whose course is 1200 leagues, if we go up as far as the Lake near Guanuco, 30 leagues from Lima, where the Maragnon takes its source; and even reckoning from the source of the river Napo, some distance from Quito, the

---

[33] See Racolta d'autori che trattano del motto dell' acque, vol. 1, page 123.

course of the river Amazons is more than a thousand leagues.

It might be said that the course of the river St. Lawrence, in Canada, is more than 900 leagues from its mouth to the lake Ontaro, from thence to lake Huron, afterwards to the lake Alemipigo, and to the lake Assiniboils; the waters of these lakes falling one into another, and at last into St. Lawrence.

The river Mississippi more than 700 leagues long from its mouth to any of its sources, which are not remote from the lake of the Assiniboils.

The river de la Plata is more than 800 leagues long, from the source of the river Parana, which it receives.

The river Oroonoko runs more than 575 leagues, reckoning from the source of the river Caketa, near Pasto, part of which falls into the Oroonoko, and part flows also towards the river Amazons.

The river Madera, which falls into the Amazons, is more than 660 leagues.

To know nearly the quantity of water the sea receives by all the rivers which fall into it, let us suppose that one half of the globe is covered by the sea, and that the other half is land, which is nearly the fact; let us suppose also, that the mediate depth of the sea is 230 fathom. The surface of all the earth being 170,981,012 square miles; and that of the sea 85,490,506 square miles, which being multiplied by 1/4, the depth of the sea gives 21,372,626, cubical miles for the quantity of water contained in the ocean. Now, to calculate the quantity of water which the ocean receives from the rivers, let us take some great river, whose rapidity and quantity of waters are known; for example, the Po, which runs through Lombardy, and waters a tract of land 380 miles long; according to Riccioli, its breadth, before it divides into many trenches, is 100 perches of Boulogne, or 1000 feet, its depth 10 feet, and it runs four miles an hour; therefore the Po supplies the sea with 200,000 cubical perches of water in an hour, or 4 millions 800 thousand in a day; but a cubical mile contains 125 millions cubical perches; therefore 26 days is required to convey a cubical mile of water to the sea: it remains therefore only to determine the proportion between the river Po and all the rivers of the earth taken together, which is impossible to do precisely. But to know it pretty exactly, let us suppose that the quantity of water which the sea receives by the large rivers in all countries is proportional to the extent and surface of these countries, and that consequently the country watered by the Po, and other rivers which fall therein, is in the same proportion on the surface of the whole earth, as the Po is to all the rivers of the earth. Now by the most correct charts, the Po, from its source to its mouth, traverses a tract 380 miles long, and the rivers which fall therein, on each side, proceed from the springs and rivers 60 miles distant from the Po; therefore this great river, and the others it receives, waters a tract 380 miles long, and 120 miles broad, which makes 450,600 square miles, but the surface of all the dry land is 85,490,506 square miles; consequently all the water which the rivers carry to the sea, will be 1974 times greater than the quantity which the Po furnishes; but as 26 rivers equal to the Po furnish a cubical mile of water to the sea in a day, of course 1874 rivers like the Po would supply the sea with 26,308 cubical miles of water in a year, and that in the space of 812 years all the rivers would supply the sea with 21,372,626 cubical miles of water; that is to say, as much as there is in the ocean, and therefore 812 years is only required to fill it.[34]

The result of this calculation is, that the quantity of water evaporated from the sea, and which the winds convey on the earth, is about 245 lines, or from 20 to 21 inches a year, or about two thirds of a line each day: this is a very trifling evaporation even when trebled, in order to estimate the water which refalls in the sea, and which is not conveyed over the earth. Mr. Halley, in the Phil. Transactions, page 192, evidently shews, that the vapours which rise above the sea, and which the winds convey over all the earth, are sufficient to supply all the rivers in the world.

Next to the Nile the river Jordan is the most considerable in the Levant, or even in Barbary; it supplies the Dead Sea with about six million tons of water every day; all this water, and more, is raised by evaporation; for, according to Halley's calculation of 6914 tons evaporated from each mile, the Dead Sea, which is 72 miles in length by 18 broad, must every day lose near nine million tons of water, that is, not only all the water it receives from the river Jordan, but also that of the small rivers which come into it from the mountains of Moab and elsewhere; consequently there is no necessity for its communicating with any other sea by subterraneous canals.[35]

The most rapid rivers are the Tigris, the Indus, the Danube, the Yrtis, in Siberia, the Malmistra, in Silesia, &c. but, as we have already observed, the proportion of the rapidity of rivers depends upon the declivity and upon the weight and quantity of water; by examining the globe, we shall find that the Danube is much less inclined than the Po, the Rhine, or the Rhone, for the Danube has a much longer course than any of these other rivers, and falls into the Black Sea, which is higher than the Mediterranean, and perhaps more so than the ocean.

All large rivers receive many others in the extent of their course; for example, the Danube receives more than 200 rivulets and rivers; but by reckoning only such as are considerable rivers, we shall find that the Danube receives 31, the Wolga 32, the Don 5 or 6, the Nieper 19 or 20, the Duine 11 or 12; so likewise in Asia the Hoanho receives 34 or 35, the Jenisca 60, the Oby as many, the Amour about 40, the Kian, or river Nankin about 30, the Ganges upwards of 20, the Euphrates 10 or 11, &c. In Africa, the river Senegal receives upwards of 20 rivers: the Nile does not receive any rivers for upwards of 500 miles from its mouth; the last which falls therein is the Moraba, and from this place to its source it receives about 12 or 13 rivers. In America, the river Amazons receives more than 60, all of which are very considerable; the river St. Lawrence about 40, by reckoning those which fall into the lakes; the Mississippi more than 40, the Plata more than 50, &c.

There are high countries on the earth, which seem to be points of division marked by nature for the distribution of the waters. In Europe, the environs of Mount St. Goddard are one of these points; another is situate between the provinces of Belozera and Wologda, in Muscovy, from whence many rivers descend, some of which go to the White Sea, others to the Black, and some to the Caspian.

---

[34] See Keil's Examination of Burnet's Theory, page 126.
[35] See Shaw's Travels, vol. ii, page 71.

In Asia there are several, in the country of Mogul Tartary, from whence rivers flow into Nova Zembla, others to the Gulph Linchidolin, others to the sea of Corea, others to that of China: and so likewise the Little Thibet, whose waters flow towards the sea of China; the Gulph of Bengal, the Gulph of Cambay, and the Lake Aral; in America, the province of Quito; whose rivers run into the North and South Seas, and the Gulph of Mexico.

In the old continent there are about 430 rivers, which fall directly into the ocean, or into the Mediterranean and Black Seas; but in the new continent not more than 145 rivers are known, which fall directly into the sea: in this number I have comprehended only the great rivers, like the Somme in Picardy.

All these rivers carry to the sea a great quantity of mineral and saline particles, which they have washed from the different soils through which they have passed. The particles of salt, which are easily dissolved, are conveyed to the sea by the water. Some philosophers, and among the rest Halley, have pretended that the saltness of the sea proceeded only from the salts of the earth, which the rivers transport therein. Others assert, that the saltness of the sea is as ancient as the sea itself, and that this salt was created that the waters might not corrupt; but we may justly suppose that the sea is preserved from corruption by the agitations produced by the winds and tides, as much as by the salt it contains; for when put in a barrel it corrupts in a few days; and Boyle relates, that a mariner, who was becalmed for 13 days, found, at the end of that time, the water so infected, that if the calm had not ceased, the greatest part of his people would have perished. The water of the sea is also mixed with a bituminous oil, which gives it a disagreeable taste, and renders it very unhealthful. The quantity of salt contained in sea water is about a fortieth part, and is nearly equally saline throughout, at top as well as bottom, under the line, and at the Cape of Good Hope; although there are several places, as off the Mosambique Coast, where it is salter than elsewhere.[36] It is also asserted not to be so saline under the Arctic Circle, which may proceed from the amazing quantities of snow, and the great rivers which fall into those seas, and because the heat of the sun produces but little evaporation in hot climates.

Be this as it may, I conceive that the saltness of the sea is not only caused by the banks of salt at the bottom of the sea, and along the coasts, but also by the salts of the earth, which the rivers continually convey therein; and that Halley had some reason to presume that in the beginning of the world the sea had but little or no saltness; that it is become so by degrees, and in proportion as the rivers have brought salts therein; that this saltness is every day increasing, and that consequently, by computing the whole quantity of salt brought by all the rivers, we might attain the knowledge of the age of the world by the degrees of the saltness of the sea.

Divers and pearl fishers assert, according to Boyle, that the deeper they descend into the sea, the colder is the water; and that the cold is so intense at considerable depths, that they cannot remain there so long under water, but are obliged to come up again much sooner than when they descended to only a moderate one. It appeared to me that the weight of the water might be as much the cause of com-

[36] See Boyle, vol. iii. page 217.

pelling them to shorten their usual time as the intenseness of the cold, when they descend to a depth of 3 or 400 fathoms; but, in fact, divers scarcely ever descend above an hundred feet. The same author relates, that in a voyage to the East-Indies, beyond the line, at about 35 degrees south latitude, a sounding lead of 30 or 35lb weight was sunk to the depth of 400 fathoms, and that being pulled up again, it had become as cold as ice. It is also a frequent practice with mariners to cool their wine at sea by sinking their bottles to the depth of several fathoms, and they affirm the deeper the bottles are sunk, the cooler is the wine.

These circumstances might induce us to presume that the sea is salter at the bottom than at the surface; but we have testimonies which prove the contrary, founded on experiments made to fill vessels with sea water, which were not opened till they were sunk to a certain depth, and the water was found to be no salter than at the surface. There are even some places where the water at the surface is salt, and that at the bottom fresh; and this must always be the case where there are springs at the bottom of the sea, as near Goo, Ormus, and even in the sea of Naples, where there are hot springs at the bottom.

There are other places where sulphurous springs and beds of bitumen have been discovered at the bottom of the sea, and on land there are many of these springs of bitumen which run into it.

At Barbadoes there is a pure bitumen spring, which flows from the rocks into the sea: salt and bitumen, therefore, are predominant matters in the sea water: but it is also mixed with many other matters; for the taste of water is not the same in every part of the sea; besides, the agitation and the heat of the sun alters the natural taste which the sea should have; and the different colour of different seas, at different times, prove that the waters of the sea contain several kinds of matters, either which it loosens from its own bottom, or are brought thither by rivers.

Almost all countries watered by great rivers are subject to periodical inundations, those which are low, and derive their sources from a great distance, overflow the most regularly. Every person almost has heard of the inundations of the Nile, which preserves the sweetness and whiteness of its waters, though extended over a vast tract of country, and into the sea. Strabo and other ancient authors have written that it had seven mouths, but there now remain only two which are navigable; there is a third canal which descends to Alexandria, and fills the cisterns there, and a fourth which is still smaller; but as they have for a long time neglected to clean their canals, they are nearly choaked up. The ancients employed a great number of workmen and soldiers, and every year, after the inundation, they carried away the mud and sand which was in these canals. The cause of the overflowing of the Nile proceeds from the rains which fall in Ethiopia. They begin in April and do not cease till September; during the first three months, the days are serene and fair, but as soon as the sun goes down the rains begin, nor stop till it rises again, and are generally accompanied with thunder and lightning. The inundation begins in Egypt about the 17th of June; it generally increases during 40 days, and diminishes in about the same time; all the flat country of Egypt is overflowed; but this inundation is much less now than it was formerly, for Herodotus tells us, that the Nile was 100 days in swelling, and as many in abating: if this is true, we can only attribute the cause thereof to the elevation of the land, which the mud of the waters has heightened by degrees, and to the diminution of the mountains in Africa, from whence it derives its source. It is very natural to believe that these mountains have diminished, because the abundant rains which fall in these climates during half the year sweep away great quantities of sand and earth from the mountains into the valleys, from whence the torrents wash them into the Nile, which carries great part into Egypt, where it deposits them in its overflowings.

The Nile is not the only river whose inundations are regular; the river Pegu is called the *Indian Nile,* because it overflows regularly every year; it inundates the country for more than 30 leagues from its banks; and, like the Nile, leaves an abundance of mud, which so greatly fertilizes the earth, that the pasturage is excellent for cattle, and rice grows in such great abundance, that every year a number of vessels are laden with it, without leaving a scarcity in the country.[37] The Niger, or what amounts to the same, the upper part of the Senegal, likewise overflows and covers all the flat country of Nigritia; it begins nearly at the same time as the Nile, and increases also for 40 days: the river de la Plata, in Brasil, also overflows every year, and at the same time as the Nile. The Ganges, the Indus, the Euphrates, and some others, overflow annually; but all rivers have not periodical overflowings, and when inundations happen it is the effect of many causes, which combine to supply a greater quantity of water than common, and, at the same time, to retard its velocity. We have before observed, that in almost all rivers the inclination of their beds diminishes towards their mouths in an almost insensible manner; but there are some whose declivity is very sudden in some places, and forms what is termed a *cataract,* which is nothing more than a fall of water, quicker than the common current of the river. The Rhine, for example, has two cataracts, the one at Bilefield, and the other near Schafhouse: the Nile has many, and among the rest two which are very violent, and fall from a great height between two mountains; the river Wologda, in Muscovy, has also two near Ladoga; the Zaire, a river of Congo, begins by a very large cataract, which falls from the top of a mountain; but the most famous is that of Niagara, in Canada, that falls from a perpendicular height of 156 feet, like a prodigious torrent, and is more than a quarter of a mile broad: the fog, or mist, which the water makes in falling, is perceived at five miles distance, and rises as high as the clouds, forming a very beautiful rainbow when the sun shines thereon. Below this cataract there are such terrible whirlpools, that nothing can be navigated thereon for six miles distance, and above the cataract the river is much narrower than it is in the upper lands[38]. The description given of it by *Father Charlevoix* is as follows:

"My first care, when I arrived, was to visit the most beautiful cascade that is, perhaps, in nature; but I immediately discovered that Baron la Hontain was deceived so greatly, both in its height and figure, that one might reasonably imagine he had never seen it.

"It is true, that if we measure its height by the three mountains you are obliged to ascend in going to it, there is not much abatement to be made of the 600 feet,

---

[37] See Ovington's Travels, vol. ii. page 290.
[38] See Phil. Trans. Abr. vol. vi. part ii. page 119.

which the map of M. Delisse gives it, who doubtless advanced this paradox only on the credit of the Baron la Hontain, and Father Honnepin; but after I arrived at the top of the third mountain, I observed that in the space of three leagues, which I afterwards had to go to this fall of water, although you are forced sometimes to ascend, you must nevertheless descend still more, and this is what travellers do not appear to have paid proper attention to. As we can only approach the cascade on one side, nor see it but in the profile, it is not easy to measure its height by instruments: experiments have been made to do it by a long cord, tied to a pole, and after having often attempted this manner, it was found to be only 115 or 120 feet high; but it is impossible to ascertain whether the pole was not stopped by some projection of the rock; for although when drawn up again the end of the cord was always wet, yet that is no proof, since the water which precipitates from the mountain, flies up again in foam to a very great height: for my own part, after having considered it on every side that I could examine it to advantage, I think that we cannot allow it to be less than 140 or 150 feet.

"Its figure is that of a horse-shoe, and its circumference is about 400 paces; but exactly in its middle, it is divided by a very narrow island, about half a quarter of a league long. It is true these two parts join again; that which was on my side, and of which I could only have a side view, has several projecting points, but that which I beheld in front, appeared to be perfectly even." The Baron has also mentioned a torrent, which, if not the offspring of his own invention, must fall into some channel upon the melting of the snow.

There is another cataract three miles from Albany, in the province of New-York, whose height is 50 feet perpendicular, and from which there arises a mist that occasions a faint rainbow.[39]

In all countries where mankind are not sufficiently numerous to form polished societies, the ground is more irregular, and the beds of rivers more extended, less equal, and often abound with cataracts. Many ages were required to render the Rhone and the Loire navigable. It is by confining waters, by directing their course, and by cleansing the bottom of rivers, that they obtain a fixed and regular course; in all countries thinly inhabited Nature is rude, and often deformed.

There are rivers which lose themselves in the sands, and others which seem to precipitate into the bowels of the earth: the Guadalquiver in Spain, the river Gottenburg in Sweden, and the Rhine itself, lose themselves in the earth. It is asserted, that in the west part of the island of St. Domingo there is a mountain of a considerable height, at the foot of which are many caverns, into which the rivers and rivulets fall with so much noise, as to be heard at the distance of seven or eight leagues.[40]

The number of rivers which lose themselves in the earth is very few, and there is no appearance that they descend very low; it is more probable that they lose themselves, like the Rhine, by dividing among the quantity of sand; this is very common to small rivers that run through dry and sandy soils, of which we have

[39] Phil. Trans. vol. vi. part ii. page 19.
[40] See Varenii Geograph. gen. page 48.

several examples in Africa, Persia, Arabia, &c.

The rivers of the north transport into the sea prodigious quantities of ice, which accumulating, form those enormous masses so destructive to mariners. These masses are the most abundant in the Strait of Waigat, which is entirely frozen over the greatest part of the year, and are formed by the great flakes which the river Oby almost continually brings there; they attach themselves along the coasts, and heap up to a considerable height on both sides, but the middle of the strait is the last part which freezes, and where the ice is the lowest. When the wind ceases to blow from the North, and comes in the direction of the Strait, the ice begins to thaw and break in the middle; afterwards it loosens from the sides in great masses, which are carried into the high sea. The wind, which all winter blows from the north over the frozen countries of Nova Zembla, renders the country watered by the Oby, and all Siberia, so cold, that even at Tobolski, which is in the 57th degree, there are no fruit trees, while at Sweden, Stockholm, and even in higher latitudes, there are both fruit trees and pulse. This difference does not proceed, as it has been thought, from the sea of Lapland being warmer than the Straits; nor from the land of Nova Zembla being colder than Lapland; but solely from the Baltic, and the Gulph of Bothnia, tempering the rigour of the north winds, whereas in Siberia there is nothing that can temperate the cold. It is a fact founded on experience, that it is never so cold on the sea coasts as in the inland parts of a country. There are plants which stand the winter in London exposed to the open air, that cannot be preserved at Paris; and Siberia, which is a vast continent, is for this reason colder than Sweden, which is surrounded on all sides by the sea.

The coldest country in the world is Spitzbergen: it lies in the 78th degree of north latitude, and is entirely formed of small peaked mountains; these mountains are composed of gravel, and flat stones somewhat like slate, heaped one on the other; which, it is affirmed by navigators, are raised by the wind, and increase so quick, that new ones are discovered every year. The rein-deer is the only animal seen here, which feeds on a short grass and moss. On the top of these little mountains, and at more than a mile from the sea, the mast of a ship was found with a pully fastened to one of its ends, which gives room to suppose that the sea once covered the tops of these mountains, and that this country is but of modern date; it is uninhabited, and uninhabitable; the soil of these small mountains has no consistence, but is loose, and so cold and penetrating a vapour strikes from it, that it is impossible to remain any length of time thereon.

The vessels which go to Spitzbergen for the whale fishery, arrive there early in the month of July, and take their departure from it about the 15th of August, the ice preventing them from entering the sea earlier, or quiting it after. Prodigious pieces of ice, 60, 70, and 80 fathoms thick are seen there, and there are some parts of it where the sea appears frozen to the very bottom[41]: this ice, which is so high above the level of the sea, is as clear and transparent as glass.

There is also much ice in the seas of North America, as in Ascension Bay, in

[41] In contradiction to this idea it is now a generally received opinion, that the mountains of ice in the North and South Seas are exactly the same depth under as they are height above the surface of the water.

the Straits of Hudson, Cumberland, Davis, Forbishers, &c. Robert Lade asserts that the mountains of Friezeland are entirely covered with snow, and its coasts with ice, like a bulwark, which prevents any approaching them. "It is, says he, very remarkable, that in this sea we meet with islands of ice more than half a mile round, extremely high, and 70 or 80 fathoms deep; this ice, which is sweet, is perhaps formed in the rivers or straits of the neighbouring lands, &c. These islands or mountains of ice are so moveable, that in stormy weather they follow the track of a ship, as if they were drawn along in the same furrow by a rope. There are some of them tower so high above the water, as to surpass the tops of the masts of the largest vessels."[42]

In the collection of voyages made for the service of the Dutch East India Company, we meet with the following account of the ice at Nova Zembla:—"At Cape Troost the weather was so foggy as to oblige us to moor the vessel to a mountain of ice, which was 36 fathoms deep in the water, and about 16 fathoms out of it.

"On the 10th of August the ice dividing, it began to float, and then we observed that the large piece of ice, to which the ship had been moored, touched the bottom, as all the others passing by struck against without moving it. We then began to fear being inclosed between the ice, that we should either be frozen in or crushed to pieces, and therefore endeavoured to avoid the danger by attempting to get into another latitude, in doing of which the vessel was forced through the floating ice, which made a tremendous noise, and seemingly to a great distance; at length we moored to another mountain, for the purpose of remaining there that night.

"During the first watch the ice began to split with an inexpressible noise, and the ship keeping to the current, in which the ice was now floating, we were obliged to cut the cable to avoid it; we reckoned more than 400 large mountains of ice, which were 10 fathoms under and appeared more than 2 fathoms above water.

"We afterwards moored the vessel to another mountain of ice, which reached above 6 fathoms under water. As soon as we were fixed we perceived another piece beyond us, which terminated in a point, and went to the bottom of the sea; we advanced towards it, and found it 20 fathoms under water, and 12 above the surface.

"The 11th we reached another large shelve of ice, 18 fathoms under water, and 10 above it.

"The 21st the Dutch got pretty far in among the ice, and remained there the whole night; the next morning they moored their vessel to a large bank of ice, which they ascended, and considered as a very singular phenomenon, that its top was covered with earth, and they found near 40 eggs thereon. The colour was not the common colour of ice, but a fine sky blue. Those who were on it had various conjectures from this circumstance, some contending it was an effect of the ice, while others maintained it to be a mass of frozen earth. It was about eighteen fathoms under water, and ten above."[43]

---

[42] See the Voyages of Lade, vol. ii, page 305, &.

[43] Voyage of the Dutch to the North, vol. 1, 3. Page 49.

Wafer relates, that near Terra del Fuega he met with many high floating pieces of ice, which he at first mistook for islands. Some appeared a mile or two in length, and the largest not less than 4 or 500 feet above the water.

All this ice, as I have observed in the sixth article, was brought thither by the rivers; the ice in the sea of Nova Zembla, and the Straits of Waigat come from the Oby, and perhaps from Jenisca, and other great rivers of Siberia and Tartary; that in Hudson's Straits, from Ascension Bay, into which many of the North American rivers fall; that of Terra del Fuega, from the southern continent. If there are less on the North coasts of Lapland, than on those of Siberia, and the Straits of Waigat, it is because all the rivers of Lapland fall into the Gulph of Bothnia, and none go into the northern sea. The ice may also be formed in the straits, where the tides swell much higher than in the open sea, and where, consequently, the ice that is at the surface may heap up and form those mountains, which are several fathoms high; but with respect to those which are 4 or 500 feet high, they appear to be formed on high coasts; and I imagine that when the snow which covers the tops of these coasts melts, the water flows on the flakes of ice, and being frozen thereon, thus increases the size of the first until it comes to that amazing height. That afterwards, in a warm summer, these hills of ice loosen from the coasts by the action of the wind and motion of the sea, or perhaps even by their own weight, and are driven as the wind directs, so that they at length may arrive into temperate climates before they are entirely melted.

END OF THE FIRST VOLUME.

Lector House believes that a society develops through a two-fold approach of continuous learning and adaptation, which is derived from the study of classic literary works spread across the historic timeline of literature records. Therefore, we aim at reviving, repairing and redeveloping all those inaccessible or damaged but historically as well as culturally important literature across subjects so that the future generations may have an opportunity to study and learn from past works to embark upon a journey of creating a better future.

This book is a result of an effort made by Lector House towards making a contribution to the preservation and repair of original ancient works which might hold historical significance to the approach of continuous learning across subjects.

HAPPY READING & LEARNING!

LECTOR HOUSE LLP
E-MAIL: lectorpublishing@gmail.com

www.ingramcontent.com/pod-product-compliance
Lightning Source LLC
Chambersburg PA
CBHW051455130726
47987CB00005B/2333